Essentials of Statistics for Research

Essentials of Statistics for Research offers a working introduction to essential statistical methods through an accessible conceptual approach without excessive mathematical details. The book emphasizes the importance of good judgment when choosing analysis approaches and illustrates the statistical analysis process through numerous examples. At its core, this text demonstrates how analysis should serve science and illuminate the stories contained within data.

Key Features

- Provides conceptual foundations of a practitioner's statistical toolkit, focusing on the role of normality, hypothesis tests, and confidence intervals
- Presents regression methods as core analytical tools while also covering t-based methods for comparing means among groups
- Demonstrates how logarithmic transformations capture relativity in relationships (such as exponential increase) rather than simply meeting statistical assumptions
- Includes over 100 graphs and visual representations to enhance understanding of statistical concepts
- Written in an engaging first-person voice that positions the authors as fellow learners alongside the reader
- Emphasizes stories, examples, and practical applications over abstract theory

This book is designed for researchers across a broad range of disciplines, from graduate students beginning their research journey to experienced professionals seeking a refresher on statistical methods. The accessible approach makes it particularly valuable for those who need to understand and apply statistical concepts without getting lost in mathematical complexity. Readers will gain practical knowledge they can immediately apply to their own research questions and data analysis challenges.

Essentials of Statistics for Research

Ken Gerow and Jorge A. Navarro Alberto

CRC Press
Taylor & Francis Group
Boca Raton London New York

CRC Press is an imprint of the
Taylor & Francis Group, an **informa** business

A CHAPMAN & HALL BOOK

First edition published 2026
by CRC Press
2385 NW Executive Center Drive, Suite 320, Boca Raton FL 33431

and by CRC Press
4 Park Square, Milton Park, Abingdon, Oxon, OX14 4RN

CRC Press is an imprint of Taylor & Francis Group, LLC

ISBN: 978-1-041-00393-9 (hbk)
ISBN: 978-1-041-00390-8 (pbk)
ISBN: 978-1-003-60960-5 (ebk)

DOI: 10.1201/9781003609605

Typeset in Palatino
by KnowledgeWorks Global Ltd.

Contents

Part IV Methods for Comparing Groups and an Approach to Repeated Measures

Preface

"The goal of any statistical analysis is to be wrong in the best possible way," Tyler Johnson.

Tyler was a grad student of statistics under Ken when he coined this aphorism, and we immediately fell in love with it. It is an edgy variation of an aphorism well-known to statisticians, "All models are wrong; some are useful," usually attributed to British statistician George Box. We have shared the "All models are wrong..." version with many students over the years; their response has been uniformly, "Sure. Whatever." And then they carry on, having for the most part tossed it aside.

Tyler's restatement, however, hits them square between the eyes. For several years, Ken asked students to respond to this at the beginning of the semester of their first graduate-level applied statistics class. They really don't like it; after all, they have spent their whole lives being right far more often than wrong, which is how they got to grad school in the first place. The sneaky, unspoken element of Tyler's aphorism is that while you don't seek to be wrong in your statistical analyses (of course), you must accept that whatever you do will be wrong in some sense or other.

You must make choices. Do you need a simple model, helpful for explanation, but perhaps at the price of worse prediction? Or is prediction paramount, in which case, you would be willing to sacrifice simplicity? Choices. Choices that require judgment. We hope this book will be helpful in improving your ability to use statistical methods with confidence. We will reference Tyler's words many times in this book, so take another look at his aphorism.

We remember well a moment like the following when we were students. A colleague came from a biology department to pose a consulting problem for the department in a seminar. Our esteemed and learned professors were unsure of how to approach the problem. They raised this idea or that (and had others object to their plans). It appeared to be utter chaos. A month or so later, one of them presented an analysis that was eventually published by that professor and the biologist. It was an epiphany to realize that the opening moves of doing an analysis can be messy[1], with perhaps many steps that eventually are retracted. That process of discovery and detective work gets left out of the final product, which looks impeccably logical and preordained.

Over the years, we have run into many early-career scientists who are frustrated that they cannot find the "right" analysis of their study in any of the textbooks they have examined. They turn to us, sure of receiving quick and sure answers, only to be met with many questions. "Why did you collect these data? What are you trying to learn? HOW did you collect the data? Remind me again of the sample size."

Most statistics textbooks show you some combination of

1. Appropriate analyses of example data sets
2. The mathematical formulae behind analyses
3. How to implement the analyses in this or that statistical package

In this text, we will offer analytic options for example data sets, but not so much of the underlying math or programming details or menu selection options. Instead, we will try to demonstrate the process of discovering and choosing analyses, as best we can, in a setting where you cannot raise your hand to ask a question. Any analysis, even in the simplest of settings, will require you to make choices.

You will see in this text a regression modeling example where the goal is to predict the waiting time of a geyser eruption given the duration of the previous eruption. Each of the several modeling choices appears to be statistically valid. How do you choose among them? In another example (predicting volume of wood in cherry trees given tree height and trunk diameter), students have proposed over ten different models! They all adequately meet modeling assumptions, and they all do an excellent job of prediction. How do you choose among them? Jump ahead to Section 11.3 if you cannot stand the suspense.

What we want to do is demonstrate how to think about the choices you will inevitably have to make and point you to a philosophy of data analysis that is not rule-bound, but is judgment-based, with your research goals as the guiding principles. A very important first step, before you start the analysis, is to define, as clearly as you can, the purpose behind the impending analysis. If you have multiple predictors (which means you must make modeling choices), you should be clear whether your goal is to explain some relationship or to make predictions. That choice can and should be impactful in your decision-making.

Many of the data sets we use are ecological in nature, mostly due to the fact that both of us work primarily with ecologists and biologists more generally. If you are, say, someone who does psychological research, or engineering traffic safety research, or... (the list goes on and on), we believe the practical foundations discussed in this book will serve you well.

The Spirit of Our Writing

Bethan Garramon Merkle is a science communication expert at the University of Wyoming, United States. She thinks deeply about how we communicate and seeks constantly to improve scientific writing to make it more useful and relevant. She espouses the idea of scholars speaking and writing as learners and helpers of learning, not omniscient experts. It is, she says, helpful to peers and junior colleagues when we do this learning transparently, in public.

In the spirit suggested by Bethann, this text will guide you through the main points of doing statistical analyses, but in a way that does not lean too hard on equations and formulas or absolute answers. We will use the first-person voice, namely "we."

We have exercises at the end of some chapters. We are not keen on "Analyze these data and report your answer" exercises for two reasons. For some such questions, it is a matter of cranking the data through a statistics package (or worse: doing some arithmetic by hand), in which case the exercise is a matter of mechanics. Fair enough, but not that compelling to us from a learning point of view. And if we supply an answer to the question, that implies that there is some single best answer, which, for real data sets, is sometimes not very realistic. There are always choices to be made, and those choices are a matter of judgment. How you choose to analyze a certain data set needn't match how either of us would do.

We express our gratitude first and foremost to our students. They inspired and informed our approach to teaching. Puzzled looks, persistent questions, and hard work on their part constantly pressed us to become better. (We hope that some of the students from our early years have forgiven us.) We are grateful to colleagues who encouraged us over the years and gently corrected us when our ideas went sideways. Anil Gore, Carson Keeter, and Rob Erikson improved the book with their review comments and suggestions. Ken is grateful to his wife Ann McCutchan, herself an accomplished writer, for her cheerful encouragement and support. Thanks to Rob Calver, Senior Publisher for the Taylor & Francis Group, for encouraging us to write this book in the first place, and to Senior Editorial Assistant Sherry Thomas for keeping us on track and organized.

We started this Preface with an aphorism; we will end with another. "Don't let the statistical tale wag the biological dog." Try to make your statistical choices shine a light on the underlying behavior of whatever you study. We entreat you to avoid, "Oh, that analysis sounds fancy; let me try it."

The data used throughout the book are hosted at https://jnavarroresearch. wordpress.com/software/#ESR as Excel files.

Note

1 To be more upbeat, we can call it a dance: two steps forward, three to the left, two to the right, back one. Perhaps even, "and twirl your partner round and round," thrown in for good luck.

About the Authors

Ken Gerow recently retired from the University of Wyoming, United States, where, as a professor of statistics for over 30 years, he taught statistics to quantitative scientists from many disciplines. Dr. Gerow earned his PhD in biometry (minor: history and philosophy of science and technology) at Cornell University, United States. He is a coauthor of the latest edition of *Elementary Survey Sampling*. He is the author or coauthor of over 90 research articles, books, and book chapters in topics ranging from the molecular and cellular world to the visible world around us (plant, animal, and human systems). Dr. Gerow considers himself a parasitic biologist because he only publishes with other people's data.

Jorge A. Navarro Alberto, PhD, is a professor emeritus at the Autonomous University of Yucatán, México, where he specialized in ecological and environmental statistics research. He earned his PhD in statistics at the University of Otago, New Zealand. His academic career spanned more than 36 years, teaching statistics for biologists, marine biologists, and natural resource managers in Mexico, and he was also a visiting professor at the University of Wyoming. He is a coauthor of the latest edition of *Randomization, Bootstrap and Monte Carlo Methods in Biology* and the co-editor of *Introduction to Ecological Sampling*. Drs. Gerow and Navarro Alberto are coauthors (with the late Bryan Manly) on the 5th edition of *Multivariate Statistics: A Primer*.

1

Introduction

The high blast of Old Faithful Geyser has just finished; it is 11:25 a.m. You are a National Park Service interpretive ranger, standing in front of that world-famous attraction, in the Upper Geyser Basin in Yellowstone National Park, baking like a chicken potpie under a blazing midday sun. There is no wind to relieve you from the heat. As the crowd of pleased geyser fans drifts away, a gentleman rushes over to you. "We just got here and missed the eruption! We are on vacation and on a tight schedule, but we don't want to miss Old Faithful! Can you tell us when it will blast off again?"

The average time between eruptions is about 70 minutes, but the waiting time can be as short as 45 minutes or as long as 120 minutes. That is an unhappily wide interval. You are already getting plenty of heat from the sun and don't want more from an unhappy visitor to whom you gave bad advice—"bad" as in the family missing the geyser or getting sunburned and hungry while waiting too long.

You know that shorter eruptions usually lead to shorter waiting times and longer eruptions to longer waiting times. The eruption you just witnessed lasted four minutes and ten seconds. Can you use that information to refine your estimate—give the rushed tourist family a more precise interval? Is there a relationship between the duration of an eruption of Old Faithful and the subsequent waiting time that you can use to your advantage? We will study that question in this book.

Having opened the book, you are likely reading this page because that's what you do with a book: start on page one. Allow us to suggest an approach to this book to maximize its practicality and usefulness. The next four chapters provide technical background information that will likely prove useful to you in the long run. But if you want to go, skip ahead to the fun stuff, turn to Chapter 6 to learn how to solve the geyser mystery; then return to these initial chapters on an "as needed" basis.

Let the journey begin.

"The goal of any statistical analysis is to be wrong in the best possible way." (Tyler Johnson.)

We started the preface with this aphorism, but we are willing to bet two donuts (so we each get one) that you did not read the preface. Almost every statistical analysis you do will involve making choices that depend on your judgment and your analysis goals. As such, you open yourself to, "This is wrong because…" or "Do you realize you left out…" or… and so on. We hope

DOI: 10.1201/9781003609605-1

that this text will help you become more comfortable with making those choices and living with them.

This introductory chapter will tell you what is and what is not in this text, but it might help to say out loud who we see as our intended audience. This book is meant for those who need to do statistical analyses but are either relatively new to the business (i.e. graduate students in quantitative disciplines) or those who are further along in their career but would appreciate a refresher and re-grounding in the practical foundations of doing data analyses. We have learned that doing statistical analyses well requires judgment. You might know the famous aphorism[1], "Good judgment comes from experience, and experience comes from bad judgment." Our experience has shown us that, for instance, listing the assumptions behind a regression analysis and then testing students on that list is not the same as them deeply learning it. All that has happened is information transfer. We gave our students practice with it, but there is only so much that can be done over the course of a one-semester period. So, you who are 5, 10, or 15 years out from graduate school and still not feeling overly confident in your analysis skills, this book is for you.

1.1 What You Will Find in This Book

Our intention is to give you a useful introduction to the essentials of doing statistical analyses. We will give you guidance on how to do analyses, not in a rote, "step one, step two, step three, done" manner, but in a manner intended to encourage statistical thinking in addition to the how-to elements. That includes the conceptual essentials—in particular, the role of the Normal distribution and the Central Limit Theorem, and concepts, procedures, and behavior of hypothesis testing and confidence intervals. We cover standard and oft-used statistical tools for means of measured variables as well as observed proportions (of success/failure, presence/absence, and so on).

Here is a partial listing of such topics:

- One- and two-sample tools for means, deploying the t-distribution, as well as for Binomial proportions; the z (a.k.a. "standard Normal" distribution) gets used here.
- We include a chapter on inference in two samples (independent or paired) when you might be interested in *relative* change rather than an increment of change. Some of the options therein require bootstrapping, and so we give a short introduction to the concepts and elements of that great tool.

- Modeling relationships between a response variable and one or more predictors, a.k.a. regression models, is the heart of the material in this book. Coverage includes
  - Model building
  - The role of logarithmic transformations
  - Interaction between predictors
  - Multicollinearity
  - Incorporation of categorical predictors
- ANOVA[2], including
  - Multiple comparisons
  - Contrasts among treatment means
- Logistic regression: modeling specialized for the case of a categorical (e.g. yes/no, present/absent, success/failure) response variable
- Response feature analysis for repeated measures data

This book focuses on conceptual understanding of statistical tools; we don't spend much time on underlying math. We will demonstrate with supporting visuals

- how the Central Limit Theorem works
- the impact of influential data points on a statistical model
- the meaning of Goodness of Fit of a model[3]

and so on. Statistical methods are based on mathematical formulae or, in some cases, algorithms. To calculate the median of a sample of numbers, you don't apply a formula; you follow a recipe. The validity and properties of statistical methods can be explained through theorems and mathematical explication. We believe you don't need to delve very deeply into that world to use statistical methods wisely and effectively.

You likely drive a car, safely and well, we hope! You do so even though you have no idea of the algorithm that delivers gasoline from the gas tank to your engine, or power to the wheels if your car is electric. You do know, however, what happens when you press your foot on the accelerator[4]. In the first few minutes of driving a different car, you instinctively learn about the sensitivity of the accelerator in this new car.

You might not know the mechanics, but **you know how it behaves.** Similarly, for instance, one needn't know the formula for the slope of a regression or for the sample standard deviation, but you do need to know how they behave. It is true that such understanding can come from studying the underlying math, but that is not the only way forward—and not the best way forward for everyone. There are plenty of excellent textbooks that teach from that perspective; this is not one of them.

1.2 What You Won't Find in This Book

It is easy when writing a book like this to fall prey to, "For sake of complete-ness…" followed by a section (or worse: a whole chapter!) on some topic the reader might never use in their entire life[5]. We want to focus on essentials[6], topics, and content that will likely stand you in good stead, whatever steps you take into other statistical tools. In that spirit, we won't spend much time on the following topics.

- Nonparametric methods[7]
- Bayesian methods
- Multivariate methods (besides, we already have a book on that!)
- ANOVA
- Data mining and machine learning[8]

1.3 How to Use This Book

This book is organized into four sections, each with its own focus. **Section One** (Chapters –5) is a section on core concepts. Language and terminology (including a statistics/English translation dictionary), the role of the Central Limit Theorem, procedures and properties of hypothesis testing and confi-dence intervals, are all considered.

   Section Two introduces regression modeling, initially through simple lin-ear regression. We walk you through what the assumptions are (and are not, interestingly), and work through a series of examples to give you practice at interpreting the assumption-testing graphs. Logarithmic transformations are often deployed, but the reason for doing so ("the response distribution was skewed, so…") misses what we see as their true value, namely that they help capture relativity in the relationships.

   Having established some details of regression essentials, **Section Three** steps into multiple regression. There are complexities and issues here that are not part of regression modeling with a single predictor, but the methodology still rests on the main elements and concepts you can learn in that section. Topics here include

- choosing which predictors to incorporate (and why), and how to do so. In particular, we consider interactions between predictors,
- incorporating categorical predictors,
- dealing with correlations among predictors: so-called multicollinearity,

- logistic regression, which specializes in modeling a categorical response variable. A surprise (maybe): it isn't always legitimate to use it to estimate changes in the probability of an event happening.

How you approach model construction can and should rest firmly on your principal goal. Are you interested in explaining the relationship between the response and one or more predictors, or are you interested in predicting the response? If the former, simpler models are to be preferred, if the latter, complicated models might be more acceptable.

Section Four deals with methods for comparing means and Binomial proportions among groups. This includes the usual one- and two-sample procedures using the *t*- and *z*-distributions. Multiple groups (i.e. classical ANOVA) are considered, albeit somewhat briefly. We *do* look at contrasts among treatment means and address the so-called problem of multiple comparisons.

Notes

1 This particular aphorism is very deeply true and can be said in different ways. It has been attributed to several different people; perhaps we will never know to whom we owe a tip of the hat.
2 We note here that regression and ANOVA are differently dressed versions of the same underlying models. We will explore that in more detail in relevant chapters.
3 It might not be what you think.
4 This metaphor can easily be extended to include the brake pedal.
5 This is sometimes a manifestation of a chronic academic condition, namely FOLO: Fear of Leaving Out.
6 Or fundamentals, if you will.
7 Short version: we believe folks turn to them unnecessarily, at the loss of being able to make inference on what they are really interested in: means and differences in means. Nonparametric procedures mostly focus on medians. See Chapter 3, on the Central Limit Theorem, for more discussion of this.
8 These get very technical and would require education and training quite beyond what many researchers need to move their respective fields forward.

Part I

Core Concepts

There are certain concepts that underlie classical statistical methods; they include the essential role of the Central Limit Theorem and principles and procedures for doing hypothesis tests and reporting precision of estimates via confidence intervals. There are also certain language conventions (jargon, if you will) that aid in the meaningful communication of results. In addition to introducing some standard language elements, we even offer a Statistics—English translation dictionary because many statistical terms are borrowed from English but take on different meanings than those same terms in every-day English.

These next four chapters will take you through that material. Feel free, if you wish, to consider these to be reference chapters to turn to on an as-needed basis. Jump ahead to Chapter 6 if you want to get started on analyses.

DOI: 10.1201/9781003609605-2

2

Data Concepts and Terminology

We will introduce language for describing the "location" of numerical data (e.g. mean, median), measures of variation (range, variance, standard deviation), and the shape of a distribution (symmetric or skewed). It is common that folks think you should use the median instead of the mean when your data are skewed. Our view of the choice is more nuanced; we discuss that in this chapter.

Almost every analysis depends on the notion of a distribution of a statistic and "standard error." The idea that a statistic has a distribution is initially somewhat counterintuitive; we try to make it seem more reasonable. Given that a statistic has a distribution, it therefore has some standard deviation. It is common to call that standard deviation the "standard error" of the statistic.

Features:

- You might be surprised at our advice regarding the mean and median in the presence of skewed data.
- Statistical jargon borrows many terms from common parlance but imbues them with different (and pointed) definitions. Furthermore, biologists and statisticians often employ words and phrases that have different meanings, depending on which culture you are speaking from. So, we end this core chapter with a small Statistics—English translation dictionary.

There are four big ideas that you need to get your head around to do statistical analyses well. They are:

1. The notion of a distribution of a statistic and standard error.
2. The Central Limit Theorem argues that, given a sample size that is not too small, the distribution of many statistics is approximately Normal.
3. Hypothesis tests: how to set them up, evaluate the data, and make a conclusion
4. Confidence intervals as a way of reporting the precision of estimates.

The standard tools for both (3) and (4) depend on (1) and (2). The classical tools for t-tests and F-tests on the one hand, and t-based confidence intervals on the other, depend on the Normal distribution. Your data need not have a Normal distribution, but the *statistic* (a mean, difference in two means, a slope from a

DOI: 10.1201/9781003609605-3

regression model...) needs to be approximately Normal. Look at that preceding sentence carefully: it implies that the statistic has a distribution associated with it. We need to understand the meaning of that, and then answer the question, "under what circumstances is that distribution approximately Normal?" This chapter develops language and notation for describing distributions, which we will later deploy when discussing the notion of a distribution of a statistic.

When Ken teaches, he starts with regression because regression models play a central role in many research projects.[1] Then, at some point in that introduction, a question arises: "What is this standard error thing?" Followed soon thereafter by "Where does the *t*-distribution fit into the picture?" And then to understand the "test of significance" for the slope, you need to understand hypothesis tests. To do estimation formally, a confidence interval might come into play. What are they? How do they work? What are their properties? This chapter will introduce you to the terminology we use throughout; reading it will leave you, dear reader, and us common ground for our subsequent work.

2.1 Notation and General Terminology

This might be the least interesting section of the entire book to you, but it is important that when we use symbols and statistical jargon, you can be clear on their meaning. When students are asked to name a couple of things that most statistical analyses depend on, two answers come quickly:

- A random sample from some defined population, and
- Your data should have a Normal distribution.

We will discuss that second one in Chapter 3, so let's deal with the first one here. Random samples are rarer than you might think, and often that's OK. Also, trying to define the population they came from can sometimes hurt your brain. Here are some real situations and specific examples.

Much excellent science is done by researchers who use animal models as proxies for human biology. Depending on the study, it might be mice, rats, or other animals, even, sometimes, nematodes[2]. The sample of mice you get to use can be purchased from suppliers who breed and raise them, including strains with certain genetic features that might be of interest, for example, trisomy mice for researchers studying Down's syndrome. That said,

- It cannot be said that the sample is random,
- The population they come from is difficult to pin down. It could be mice of a certain strain over a specific time period, so that the concept of a parameter to be estimated gets pretty slippery, and
- Interest is in their role as an imperfect model of human biology.

Still, good science is done in this setting, so long as inferences are made with care.

Psychologists and kinesiologists often use student volunteers in their studies. So, no random samples, and from a population of people in a narrow age range, and likely also a narrow geographic and socioeconomic group. Researchers usually take care to make their formal statistical inferences appropriately.

Here is an example where random sampling was indeed used. Dauwalter *et al* (2022) studied the dynamics between brook trout and Yellowstone Cutthroat Trout in a watershed at the intersection of the Idaho, Utah, and Nevada borders. They chose their study sites using stratified random sampling. They took care to make formal inference only to their study area, i.e., the population of stream reaches from which they took a sample. Inference beyond that, which readers might wish to make, is dependent on critical thinking and imagination. "Are there features of this study area (e.g., elevation, aspect, temperatures) that differ from an area of interest to me in such a way that their findings might not apply?"

Defining the population can get very slippery. Imagine wanting to estimate the average age of registered voters in the United States. Over the time it took you to read that sentence, the population changed. Some of them died, and some others turned 18 and became registered voters. And so it goes. Studies that have a random sample from the population of interest are not overly common. It is a statistical ideal worth keeping in mind when designing your study and reporting your results. And the properties of statistical estimators, which can be described perfectly when everything is crisply defined, are still approximately fine in our real world[3].

As we move ahead, we will use a small collection of symbols (Greek letters and the like). First, we focus on a single sample from some population, in particular, commonly estimated parameters of a population. The "fathers" of modern Statistics, Karl Pearson, Sir Ronald Fisher, and William Gossett (a.k.a "Student") used Greek letters to represent parameters. However, unlike Pearson, Fisher and Gossett insisted on clearly distinguishing parameters and estimates. To do so, they suggested using Greek letters for the former and Latin letters (our usual alphabet) for the latter. For the most part, we adopt their style.

Y is the numerical variable of interest[4].

μ_Y is the population mean of Y.

σ_Y is the standard deviation (defined below) of Y; σ_Y^2 denotes the variance.

η_Y is the median of Y in a population.

Sample analogues of the foregoing parameters are

$Y_i; i = 1, 2, ..., n$: individual observations of Y in a sample of size n.

$\bar{Y}$ is the sample mean of Y, pronounced "Y bar".

S is the sample standard deviation; the sample variance is denoted as S^2.
$\tilde{Y}$ is the sample median of Y

Notes:

- We use a lower-case letter to denote an actual observation. For instance, $\bar{y} = 12.6$ would represent the mean from an actual sample, whereas $\bar{Y}$ represents the sample mean as a concept.
- 50% of the values in a population are less than the median, and 50% are greater.
- If we need to refer to an estimate of a parameter, we will use the symbol with a carat above; for example, $\hat{\mu}_Y$ represents an estimate of μ_Y.
- When the context is clear, we might omit the subscript.
- When we get to regression models, it will be convenient to use different symbols for the response variable and the predictor variables. There we will use X as the symbol for the predictor, subscript it if there are more than one.
- Numerical variables can be on the so-called "interval" scale; zero on such a scale is not a true zero (e.g., temperature in Fahrenheit or Celsius) or "ratio" scale. The latter has a true zero. Temperature in Kelvin has an absolute zero; it is a ratio scale variable.
- We may sometimes use subscripts, e.g., S_Y and S_X for the SD of Y and X, respectively.
- Ratio scale variables never go below zero[5]. Many variables of interest, especially in biological data, are ratio variables. The importance of this will become clear when we discuss *relative* changes in values because making relative comparisons makes sense for ratio variables, but not for interval variables. Forty kilograms is twice as much as 20, whereas 40°C is not twice as hot as 20°C.

2.2 We All Want a Distribution

In this section, we will describe numerical and graphical options that are commonly used to describe sets of numerical data.

2.2.1 Statistics for "Location" of Numerical Data

The sample mean is defined as $\bar{Y} := \frac{1}{n}\sum_{i=1}^{n} Y_i$. A few comments on notation are in order here.

- The symbol ": =" reads as "is defined to be"

- $\sum_{i=1}^{n} Y_i$ is shorthand for $Y_1 + Y_2 + ... + Y_n$.
- When the context is clear, we might simplify this to $\sum Y_i$.

Given a random sample of n values of Y from some population, $\bar{Y}$ is said to be unbiased for μ. It implies that $\bar{Y}$ doesn't have any systematic tendency to be less than or greater than μ. **In any one sample, it will be randomly above or below μ; it is quite unlikely to be precisely equal to it.**

Another useful summary measure for the location of a set of numbers is the median. Many statistics are calculated by a formula; the mean is an example. Others require an algorithm or recipe.[6] The algorithm for calculating the sample median is as follows. Sort the sample from smallest to largest values. If n is an odd number, the median is declared to be the middle value in that sorted list. If n is even, it is the average of the two middle values. A couple of notes…

1. If n is an odd number, clearly no single value has 50% of the sample values that are smaller and 50% that are larger. So, the 50–50 business is a little squishy. That's life for you.

2. Given an even sample size, choosing to take the average of the two middle values is a convention; *any* value that is between those two would have the property that 50% are smaller and 50% are larger.

The sample mean is used to estimate the population mean, and the sample median is used for the population median. Defining those parameters begs the question, "what is the population?" As we suggested in Section 1.1, defining the population can sometimes be tricky; indeed, sometimes the population of interest might be considered infinite. If the population can be crisply defined, and is finite with, say N elements, then the definitions we gave here for sample values of the mean and median apply. Often, this is not so, and we simply live with a vague notion of what those parameters are.

We will return to a discussion of the mean and median shortly, when we discuss shape.

2.2.2 Measures of Variation

The simplest and most intuitive measure of variation is the range (largest value minus smallest value). Unfortunately, it hasn't been so useful in creating statistical procedures for testing or estimation, and so it is not often used in statistical practice. More useful are methods that are based on the question, "How far from the sample mean are individual data values?" The difference between each value and the sample mean is called its deviation from the mean. For instance, $Y_3 - \bar{Y}$ is the third such deviation.

A good place to start might be the average of them, the "average deviation." Unfortunately, the positive deviations and negative ones cancel out perfectly, so this value will always be zero. Reboot. What about the average "size" defined by taking the absolute value? That does exist, the so-called mean absolute deviation, which has a great acronym: MAD. Not so popular in use, though. The one that we have landed on for most settings is based on squaring the deviations. This takes care of the negatives, of course, but has its own drawbacks, correction for which led to the well-known standard deviation.

The path to the standard deviation is not long but is a bit arcane. It looks like this:

1. Calculate the so-called sample variance: $S^2 = \frac{1}{n-1}\sum_{i=1}^{n}(Y_i - \bar{Y})^2$. Notes:

    a. The divisor *n*-1 is used so the statistic is an unbiased estimator of the variance in the population. Having an unbiased statistic makes sense. The formula espoused in the United Kingdom and other places divides by *n*, in which case the statistic can be called the average of the squared deviations. This also makes sense[7]. They yield numerically very similar answers, except for very small sample sizes.

    b. This statistic does not lend itself to intuition very well. Suppose for data measured in minutes, that the variance is 184.8; units are minutes squared, no less! It is difficult to develop much intuition for this value.

2. The SD is simply the square root of the variance. For the foregoing example, $s = 13.6$. The units are minutes, and it is easier to understand their meaning, as we will now show.

The SD is numerically related to the range in a loose but convenient way. The range is often close to four times the SD. Figure 2.1 shows a sample of ten such ratios drawn from data sets in our collection. The average ratio is 3.85. Our

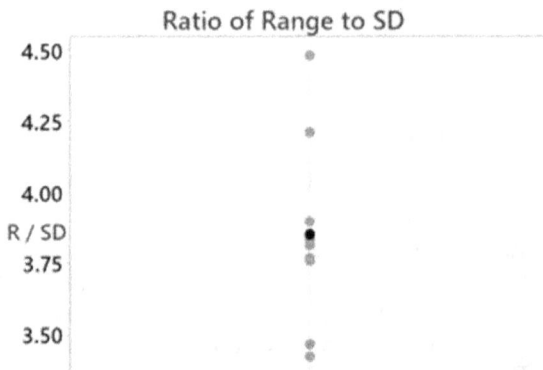

FIGURE 2.1
Ratio of sample ranges to sample SDs for ten data sets. The mean is 3.85.

point? If someone reports to you the SD from their sample, you can quickly multiply by 4 and have a good guess for the range of their data values. Having a feel for data is a good thing, and this rough relationship helps.

2.2.3 Shape of a Distribution

Distributions are commonly described as symmetric or skewed[8]. Figure 2.2 shows four examples, two symmetric and two skewed.

A question for you: which of (c) and (d) is skewed right and which is skewed left? The question is interesting to us because we have learned that when naive (but very smart) students are asked that question, the majority will declare panel (c) to be skewed left, and panel (d) to be skewed right[9].

Karl Pearson was one of the founders of modern statistical methods; in fact, he founded one of the world's first university departments of statistics in 1911. He came up with a formula for what he called skewness. A symmetric distribution yields a skewness coefficient of zero. Since we usually orient a number line so that positive values are to the right of zero, and negative values to the left, he called distributions that yielded a positive skewness "skewed right," and conversely, distributions with negative values he called skewed left. Unfortunately, that is counter to how we use the term in everyday English. If, for instance, we say that car colors skew white

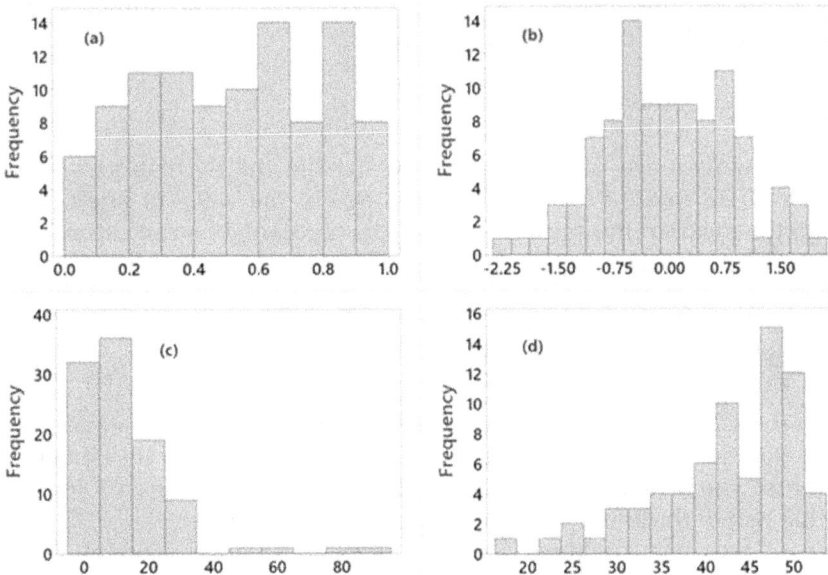

FIGURE 2.2
Two symmetric distributions: uniform (a), Normal (b), and two skewed distributions: exponential (c) and mouse lifespan data (d) (Weindruch et al, 1986).

in the southern United States, we imply that more cars there are white. When transferring that sensibility to a histogram, it makes perfect sense to declare the skew to be in the direction where the bulk of the data lies: panel (c) that would be to the left and the one in (d) to the right. Alas. Tis not so. See the Statistics-English translation dictionary for a picture...

2.3 Your Data are Skewed: Do You Use Means or Medians?

Given that the data are skewed, should you use the sample mean or median for inference? When Ken asks students in his classes to weigh in on the matter, they tend to vote in favor of the median, pointing out that "the mean is sensitive to outliers," while the median is robust to them.[10] While that statement is true, it is not a failing of the mean; indeed, it can be a strength, which we illustrate with the following example.

Suppose you have a random sample of house prices in Laramie, WY. It won't surprise you to imagine the histogram of the data being skewed right. A large percentage of the values will be between $300,000 and $500,000, with predictably fewer in the $500,000–$700,000 range, and so on. Finally, way off to the right, all by its lonesome, will be Ken's house[11].

The mean and median will be meaningfully different, for sure. We will argue here that the choice between mean and median depends mainly on the purpose for which the data were gathered.

Case 1. Civil servant establishing tax rates.

Homeowners are taxed to generate income for the county's school district. The amount is a small percentage of the value of one's house. From the mean in the sample (but not the median), one can estimate the total real estate value in the county: multiply the mean by the number of houses.

Case 2. Prospective house buyer.

On the other hand, suppose we are looking into the Laramie real estate market to decide if we want to purchase a house as a rental investment. In that case, the median is an attractive statistic, since it represents "typical" in that half of the house values are smaller, half are larger.

These days, one would not use a sample of house prices, since all the relevant data are in a computer database. Our point is that which statistic to use is a function of your purpose in doing the study in the first place, not a consequence of the shape of the distribution of the data. Are you interested in the mean in the population or the median? *That* should drive your choice. In scientific studies, attention is on the total, often only implicitly. For example, if you are measuring the weights of fish in a sample from some population, the average allows you to at least implicitly understand the totality of fish biomass in the population; you can't get there from the median.

We also note that the distribution being skewed does not preclude using methods based on Normality, thanks to the Central Limit Theorem; see Chapter 3. Given a sufficiently large sample, the distribution of the sample mean will be approximately Normal, usually true even if the population from which it came is not.

2.4 Graphical Representation of Data

The goal of a graphical representation is to adequately summarize the data in a visual manner. The two most used tools are histograms and boxplots; the histogram is perhaps foremost. Figure 2.2 uses histograms. The range of the data is divided into a suitable number of "bins" and the graph represents by a bar how many data points fall into each bin. How many bins to use and what the "breakpoints" should be is a matter of choice, and decades ago, it was trial and error. These days, statistics packages will do that for you automatically, since we have settled on general patterns that seem to serve well enough.

Making histograms was tedious before computers and statistics packages came along. John Tukey (Tukey, 1977) invented a simple graphical summary he called a boxplot, or box and whisker plot. Figure 2.3 shows a boxplot of the mouse lifespan data from Figure 2.2d.

Notes: John was seeking a graph that could be drawn quickly and easily with pencil and paper, yet would convey essential information about the data. The box captures the middle 50% of the data. The line inside shows the median of the data. The fact of it being off-center to the right is due to the data being skewed left. The sides of the box are the 25th percentile (first quartile) and 75th percentile (third quartile), respectively[12]. Then he added a line to extend to the minimum and maximum values, the so-called whiskers. While contemplating his creation, he realized that the line visually implied the presence of data all along its length. A large outlier (very small or very large value) might mislead the reader. Accordingly, he came up with a rule: draw the whisker no longer than 1.5 times the width of the box. If there are

FIGURE 2.3
Boxplot of mouse lifespan data.

any data points further away, indicate them with a special symbol[13]. We used a black diamond in Figure 2.3. Your statistics package might use different symbols.

Typically, a boxplot conveys less information about the shape of the distribution than does a histogram. Once histograms became available with the click of a mouse, they became the favorite one to use. Boxplots are useful when details are not required. For example, if many distributions are to be compared, the simplified summary of the box plot might be adequate.

Given a small data set, one could show an individual value plot. Figure 2.1 is an illustration. Since most studies don't use very small sample sizes, these plots are not often seen.

2.5 A Translation Dictionary of Statistical Terms

This short dictionary focuses on words that have different meanings in everyday English, or, in some cases, in bioscience lingo, than they do in their technical use when doing statistics. Such terms can be confusing when the speaker or writer uses them technically, but their listener or reader uses them otherwise.

2.5.1 Average (also Called the "Mean")

- **Statistical usage:** It references a simple formula: Add up all the numbers in a sample, divide by the number of numbers. It is a measure of "location" of the data set.

- **Everyday English:** It means "typical" or "usual". In Las Vegas, they take advantage of this by reporting that (say) on average, 95% of all dollars gambled are returned to the gamblers. The problem is, the typical return is $0.00 (the first quartile, the median, the third quartile... all $0.00), while a very, very few walk out with lights flashing, bells clanging, and a large cart to carry all their money. The average return rate might be their advertised percentage, but the typical return rate is about zero.

2.5.2 Bias

- **Statistical usage:** A statistic is said to be biased if it has a systematic tendency to fall below or above the parameter being estimated. Unbiased, of course, implies no such tendency. Due to our everyday use of the word, you might assume that biased statistics are *ipso facto* bad and to be avoided. As counterexamples, the sample standard deviation and sample correlation coefficients (from relevant random samples) are both biased downward: they have a tendency to systematically underestimate the corresponding population parameter. We don't often dwell on that feature because the bias in each case is usually very small and shrinks with increasing sample size.

- **Everyday English:** To be accused of being biased is usually not a good thing. This definition captures it: prejudice in favor of or against one thing, person, or group compared with another, usually in a way considered to be unfair. This notion, that being biased is likely a bad thing, might affect your initial response to being introduced to a biased statistic.

2.5.3 Census

- **Statistical usage:** In a sampling setting, a census implies you have measured the response variable on ALL of the units in a population. Therefore, precision has attained perfection: the standard error will be zero.
- **Biologists' usage:** Biologists use the word "census" to mean "count," as in, "We censused caribou on each plot." Here, the sampling units are plots, of which they have selected a sample from the population. On each plot, they measured their chosen response, namely the number of caribou.

2.5.4 Confidence

- **Statistical usage:** When constructing a confidence interval, the "confidence level," commonly chosen to be 95%, dictates how wide the interval will be. By way of reminder, this leads to statements such as, "We are 95% confident the true mean waiting time is between 68 and 74.5 minutes." Choosing a higher confidence level leads to a wider interval, and a lower level to a narrower one.
- **Everyday English:** You know the meaning… One might be highly confident in your data for a variety of reasons, but this will not make a confidence interval wider or narrower.

2.5.5 Deviation

- **Statistical usage:** Specifically, the subtraction of one thing from another… in a regression model, for example, a residual measures the deviation of one response value from the fitted line. It would also apply to the difference between a single datum and the mean of a sample.
- **Everyday English:** The word is used similarly, but more broadly: to "deviate from the norm" is simply to be different. Someone who does so might be called "deviant" rather than a "deviation."

2.5.6 Goodness of Fit

- **Statistical usage:** In short, testing for goodness of fit of a model is asking with the given predictors already in the model, whether something more complex (e.g., a squared term, an interaction term, or a log-transformation) might be required.

- **Everyday English:** We don't use that phrase, *per se*, but when people first see that phrase, an intuitively attractive misunderstanding gets in the way of the true meaning. In particular, people might think that a model with low p-values or decent R^2 are evidence of a good fit. It is true that these usually go along with a good fit, but they are not definitive. See Chapter 6 for details.

2.5.7 "In General"

- **Mathematical usage:** When a mathematician says, "in general," they mean, "always, every time (and I can prove it)."

- **Everyday English:** Here, the phrase means, "most of the time," or "usually." If you see that phrase in this text (mathematicians beware), we are using it as an everyday English phrase.

2.5.8 Interaction

- **Statistical usage:** An interaction between two predictors in a multiple regression implies that the effect, as measured by its slope coefficient, of one predictor changes depending on the numerical value of the other, and vice versa. This has no bearing on whether the two are correlated. Two predictors that are independent of one another might have an interaction; two that are correlated might not.

- **Everyday English:** An interaction between two people implies a relationship of some kind. Take the word "relationship" back to statistics; a relationship between two predictors implies they are correlated. This constitutes a publishable error. We have seen research articles that stated, "We tested for interactions among the predictors by examining correlations among them." Oops.

2.5.9 Linear

This word appears as an adjective in the phrase, "linear model," and the meaning is not what it appears to be to an English speaker

- **Statistical usage:** A linear model is a model formed from a linear combination of the model parameters, a perhaps arcane mathematical concept. For the sake of brevity and lack of context here, we will reserve a deeper explanation until Chapter 8.

- **Everyday English:** "Linear model" must mean the model is a straight line. What else could it be?

2.5.10 Mean

- **Statistical usage:** See "average." It has several uses in everyday English, so in this book we will say, "sample mean" or "population mean."
- **Everyday English:** For example, if you interrupt other people when you are supposed to be having a civilized conversation, you are being mean. It is a synonym for "intend" ("what I mean to say is...") and "imply" (What this means is...").

2.5.11 Normal

- **Statistical usage:** This is the name for a certain, very important distribution in statistical practice. It is symmetric and bell-shaped,[14] and it is the distribution associated with very many commonly used statistics[15]: for instance, the mean from a random sample, the difference between two means, the slope from a regression model.
- **Everyday English:** The word simply means "usual" or "typical." We note in passing that since very many variables we study have skewed distributions, it might be fair to say that it is not normal to have a Normal distribution for your data.

2.5.12 Population

- **Statistical usage:** A population is the collection of things (e.g., people, plots of land...) from which a sample is selected. A biologist studying pronghorn in a certain region of Wyoming might have a sample of 35 one-square-mile plots, from a large number of such plots, on which he/she counts pronghorn. Let's push on this a little more. It is convenient to declare that the statistical population is a population of *numbers* (abundance of pronghorn here) from which you had taken a sample of such numbers. From that perspective, it makes sense to talk about your sample mean and the population mean of those values.
- **Biologists' usage:** The biologist might say he/she is studying the population of pronghorn in a certain region of Wyoming. Both uses of the term are reasonable, but it behooves you to pay attention to the cultural context. Sometimes we have found ourselves talking past a biologist (and vice versa) because we use that same word but with different meanings.

2.5.13 Random

- **Statistical usage:** In reference to selecting elements[16] into a study, random selection implies that each element of the population has a specified chance[17] of being selected. Interestingly, random selection of elements is not necessary in a regression model setting. In fact, *choosing* values of the predictor is often smart[18]. This is counterintuitive, but examples are easy to imagine. In a growth chamber study of the effect of temperature and watering levels, the researcher would most likely choose temperature and watering settings. In a regression or ANOVA, the required randomness in the data is met by assuming that Y responds randomly to levels of the predictors.
- **Everyday English:** Something that happens randomly happens haphazardly, with no control over the chance. Even more extremely, the word now often is taken to imply "weird" or "unusual." A person spills their beverage on their shirt. "That's random…" Be aware of which language you are speaking…

2.5.14 Significant

- **Statistical usage:** In reference to a statistical hypothesis test, a "significant result" simply implies that one would reject the null hypothesis, that the p-value is less than alpha. It does *not* convey any information about whether the effect being tested is large or small, only that it appears to be real, not an artifact of randomness.
- **Everyday English:** It means, "important," of course. Two averages might be "significantly different" statistically, but the difference might or might not be important. That designation is a judgment call on the part of the researchers.

2.5.15 Skewed

- **Statistical usage:** The term is used to connote a distribution that is not symmetrical. A distribution may be left-skewed or right-skewed. The definition of left-skewed and right-skewed is not intuitive, hence the need for this dictionary entry. A statistician would say the data in Figure 2.4 are skewed right.
- **Everyday English:** When we say that the results were skewed in a certain direction, we mean that more of them went in that direction than in the other direction. An English speaker looking at Figure 2.4 might declare these data to be skewed to the left, since that is where the bulk of values lie.

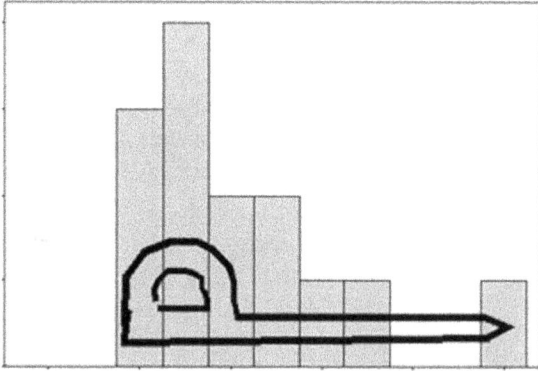

FIGURE 2.4
A visual mnemonic to help remember the definition of skew. Here, the human skewer is pointing to the right, so these data are right-skewed.

2.5.16 Standard

- **Statistical usage:** It is most often used as an adjective to describe a distribution, as in "standard Normal distribution." The distribution has been rescaled to have a mean of zero and standard deviation equal to one. Having such a named distribution was especially useful for creating tables to deduce probabilities for data presumed to have a Normal distribution. And this in turn was especially useful for teachers of introductory statistics, who could ask their students, "What is the probability that Y lies between 7 and 10, given that it comes from a Normal distribution with mean 5 and SD 4?" You would note that 7 is $\frac{(7-5)}{4} = 0.5$ SDs above the mean, and that $\frac{(10-5)}{4} = 1.25$ SDs above, and could translate the problem into, "What is the probability that Z, which has a standard Normal distribution, is between 0.5 and 1.25?" And then you could look up the answer in a table. It *was* cool since you could apply the method to *any* Normal distribution; all you needed to know was the mean and the SD. And then you passed your Intro Stat class and most likely never saw this use again in your life.
- **Everyday English:** It means "typical" or "usual." See "average" for another such word.

2.5.17 Standardize

- **Statistical usage:** Sometimes you read or hear, "I standardized my variables, ..." and then they go on. If you ever hear anyone say this, ask, "Specifically, what did you do?" Then ask, "Why did you do that?" You need to ask the first question because there is no one definition of the statistical verb, "to standardize," and the answer to the second question is, well, see the Everyday English meaning of "standard" just above. Some folks use the word "normalize" in this context. It similarly has no single definition. It is not a bad thing to do, *per se*, but the "what" and "why" should be considered explicitly.

- **Everyday English:** We have run into people who standardized their data because they heard that someone else had done it, and, by golly, they are going to do the standard thing, too, so they don't get into trouble.

2.6 Chapter Summary

We introduced language for describing features of collections of numbers. If the collection is a sample, these values are called statistics; if it is a population, we call them parameters. The mean and median are the two most used measures of the "location" of a set of numbers. The distribution of a set of numbers can be either symmetric or skewed; if the latter, then the mean and median are noticeably different numerically. For skewed data, your purpose should dictate whether you favor the mean (use this if you wish to imply "total") or the median, which might better represent "typical."

The range of values in a sample (maximum minus minimum) is the simplest measure of variation. Early on, though, it fell short in terms of utility for developing statistical procedures. The most used measures in classical methods are the variance and its square root, called the standard deviation. The SD and range have a predictable, albeit somewhat loose connection: the range is often three or four SDs. For quite large samples, it might be four or five.

Graphical summary tools include histograms, boxplots, and individual value plots. Before computers made histograms a snap, boxplots were quite popular. They could be made quickly with a pencil and paper. They visually demarked the minimum, maximum, median, and first and third quartiles of a data set. The two quartiles bounded the middle 50% of the data values and provided another measure of variation: the interquartile range. For small data sets, individual value plots are sometimes used, but they get pretty cluttered if the data set is large. These days, most folks lean toward histograms as a compromise between the other two. They offer more detail than a boxplot and a greater summary than the individual value plot.

There are several terms used in statistics that have a different meaning in everyday English (or in some cases a different meaning if used by a biologist) than they do in statistics. You need to be cognizant of the cultural context to be sure of the meaning of these words. We have provided a statistics/English translation dictionary that provides clarity on these terms.

2.7 Exercises

1. **Choose (a) or (b) and explain your answer.** Suppose you see a summary of a ratio-scale variable, and for it, the SD is large relative to the mean. For example, the mean might be 20, and the SD is 15. Are these data

   a. symmetric or skewed (and if skewed, in which direction), or

   b. you cannot tell the shape from this information?

2. Consider the following statement. "In a large sample of statisticians, 50% of them have a below-average IQ." Does this tell you anything about the shape of the distribution?

3. Suppose you know that the mean in a sample is quite less than the median. What does this imply for the shape of the distribution?

2.8 Solutions to Exercises

1. Given that we have ratio-scale data, we know that there will be no values less than zero. In this chapter, you learned that the range of data values is on the order of three or four times the SD. So, here, the range is about 50 or so. Data must stop at 0, which is 1 1/3 SDs below the mean, but can stretch out above the mean. This in turn implies distribution is likely to be skewed right.

2. Since 50% of the values in a data set are below the median, this sentence implies that the mean and median are equal. In that case, it implies that the distribution of IQs among statisticians is approximately symmetric.

3. When the mean and median noticeably differ, you could deduce that the distribution is skewed. The mean is usually off in the direction of said skew, so these data are likely skewed left. Lifetime data often have this feature, which makes them unusual among biological variables.

Notes

1 You might be thinking in terms of treatment groups, and so ANOVA is on your mind. As we shall see, ANOVA and regression are so intimately related they could be considered different flavors of the same idea.

2 Most nematodes are mostly very tiny (on the order of a millimeter or two), hardy, worms. In particular, *Caenorhabditis elegans* is easy to maintain in laboratories and has a short lifespan and a small genome, making it highly useful

for research. Fun fact: *Placentonema gigantissima* is a nematode that can attain a length of over 8 meters; it parasitizes the placenta of the sperm whale.

3   In some sense, here we run into Tyler's aphorism for the first time. Statistical analyses will never be perfect, so in some sense or other are doomed to be wrong, but we do the best we can. It's like real life.

4   Data can also be categorical. We will discuss those in a later chapter.

5   Some things we measure can indeed take on a value of zero (e.g. when counting things). In that case, the defining property of the ratio scale take a bit of a blow. Most often this poses no difficulties in an analysis, but it does get troublesome when taking logarithms. We will discuss that in the relevant chapter.

6   The word algorithm comes to use from the name of a Persian scientist and mathematician who lived around 800 AD. Our words algebra and alchemy come from the same source.

7   We run smack into Tyler Johnson's aphorism again. Both forms have some appeal; neither is perfect.

8   There are other valid features (e.g. kurtosis), that are not often necessary to think about when analyzing data, so we will skip those here. We intend to help you with most of the ideas and tools you will need most of the time.

9   Naive meaning they had not been exposed to the relevant definitions

10   Bad mean!

11   Ha! Fat chance.

12   It would not be wrong to call the median the 50th percentile; further, you can think of the first quartile as the median of the lower half of the data values and the third as the median of the upper half.

13   This highlighting of outliers is a very handy feature of boxplots.

14   That is not its defining feature: other distributions could be similarly described.

15   This is often true provided the sample size is large enough. See Chapter 3 on the Central Limit Theorem for details.

16   An element could be a person, a plot of land, an animal, or a plant…

17   In a simple random sample, each element would have an equal chance, and selection of any one element does not affect the chance associated with any other element.

18   If you choose values to include very low and very high levels of the predictor, you will have a better chance of pinning down any relationship.

References

Dauwalter, D.C, M.A. Baker, S.M. Baker, R. Lee, and J.D. Walrath. 2022. Physical habitat complexity partially offsets the negative effect of Brook Trout on Yellowstone Cutthroat Trout in the peripheral Goose Creek subbasin. Western North American Naturalist 82(4): 660–676

Tukey, John W. 1977. Exploratory Data Analysis. Addison-Wesley. 720 pages.

Weindruch, R., R.L. Walford, S. Fligiel, and D. Guthrie. 1986. The retardation of aging in mice by dietary restriction: longevity, cancer, immunity and lifetime energy intake. Journal of Nutrition 11(4): 641–654.

3

The Statistical Law of Gravity
(a.k.a. the Central Limit Theorem)

It is perhaps counterintuitive to consider that the statistic you are studying (a single mean, difference between two means, and so on) has a distribution. It does, though, in a certain sense, and understanding the meaning of that is core to understanding why classical statistical methods work. We will explain that to you and share the language to discuss that distribution.

An essential feature of the distribution of a statistic is that, for many commonly used statistics, that distribution is approximately Normal, despite that the data behind it are usually not. We illustrate how the Central Limit Theorem (CLT) does its work to make that happen.

Speaking of counterintuitive, after all that work to understand that your statistic has a Normal distribution, we promptly toss it away for many statistics and use the t distribution instead. We will show you what drove William Gossett into discovering that distribution.

All classical statistics tools using the t, z, and F distributions require a Normal distribution. This is *not* a requirement of your data, per se, but rather it pertains to the distribution of the statistic you are using. We will introduce the notion of a distribution of your statistic and then step into the CLT. In short, the CLT says that the distribution of certain statistics will be approximately Normal given a sufficient sample size[1]. A partial list of statistics to which this applies includes:

- A sample average
- The difference between two averages
- A sample proportion of some outcome of interest[2]
- The difference between two proportions
- Intercept and slope coefficients from classical regression models

3.1 The Distribution of a Statistic and the Meaning of "Standard Error"

If we were to flip a fair coin and ask you, "what is the chance it came up heads?," most of you would answer, "50%" or "50–50" or the like. In so doing, you instinctively conjured up a very large number of such coin flips, of

DOI: 10.1201/9781003609605-4

which approximately 50% would be heads, and applied that understanding to answer the current question[3].

Strictly speaking, your sample mean does not have a distribution. To think about the distribution of the sample mean, or any other statistic arising from a sample, requires the same act of imagination that you instinctively used with the coin flip. You understand that if you were to repeat the study, you would not get precisely the same data, and therefore would get a different value for the mean. Imagine repeating the experiment a very large number of times in the blink of an eye. Imagine that every time you do, all you do is write down the average. When you are done, you will have a distribution of that statistic[4].

Figure 3.1 illustrates this idea by way of simulation. In a study in a book by Härdle (1991) of waiting times between eruptions of Old Faithful geyser[5] in Yellowstone National Park, $n = 272$ waiting times were recorded (Figure 3.1a). They range from 43 to 95 minutes; the average was 70.90 minutes. Using a resampling technique called bootstrapping, we simulated repeating the study 1000 times.[6] Each repeat uses 272 values, drawn in such a way that for each draw, every datum in the sample has an equal chance of being picked, but a chosen value is not removed from the list, so it could be drawn a second or even a third time. The resulting distribution of the means is Figure 3.1b. Those values range from 68.25 to 73.84 minutes, with a mean of means of 70.93. This is different from the sample mean by what is called "simulation slop"; had we used, say, 10,000 repeats, the result would have been even closer to 70.90.

As noted above, the range in the distribution of means is quite smaller $(73.84 - 68.25 = 5.59)$ than that for the actual data $(95 - 43 = 52)$. This is because each mean combines information from 272 data points, so each of them is likely to be closer to the true (assumed unknown) mean than individual observations in the sample. Note also that the distribution in the sample is bimodal: there is a smaller collection of points centered roughly

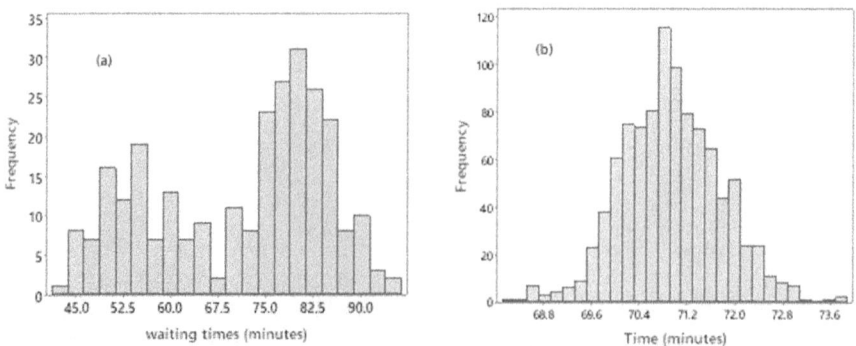

FIGURE 3.1
Histograms of a sample of waiting times between eruptions of a geyser (a) and of means (b) from 1000 simulations of repeating the study.

around 55 minutes, and a larger subset centered around 80 or so. On the other hand, the simulated distribution of the means is roughly symmetric. We will return to that point below.

The SD in the distribution of means in Figure 3.1a is 0.84, as calculated by a statistics package. But you don't need to learn how to program simulations to get this. Assuming you have a random sample from some population, the sample mean estimates μ. Recall that σ symbolizes the population SD, while s is the sample SD and n is the sample size. The SD of the distribution of the mean is $\sigma/\sqrt{n}$, conveniently estimated by $S/\sqrt{n}$. The phrase "standard error of the mean" is in fact just an unfortunate synonym for "SD of the distribution of the mean." We wish that whoever first studied this had simply called it that.

Warning: As a general notion, the formula for the standard error is not $S/\sqrt{n}$. Wait, what? Didn't you just read that it was $S/\sqrt{n}$? Asked on an exam for the formula for the standard error, many students will say $S/\sqrt{n}$. But we need to be careful, and this is not a minor point.

The behavior of a statistic arises from two things:

1. The study design, and
2. The formula or algorithm that yields the statistic in question.

To estimate how far students have to go to get from home to class, suppose we ask the five students sitting in the front row that question. The average and standard deviation of those numbers can be easily calculated. Suppose the study is repeated many times and that the sampling scheme is to keep asking those same five students. It dawns on the class that the sampling design sucks and that the numerical value of the statistic will never vary: its standard error is zero. Having an infinitely precise statistic sounds like a good thing, but here it is clearly not something to brag about.

The formula for the standard error of every statistics is not $s/\sqrt{n}$. As you will see in the rest of this text, statistics other than the mean (e.g. the difference in two means, or the slope in a regression) each have their own standard error formula.

3.2 The Distribution of the Mean is Approximately Normal (under Certain Conditions)

As we have learned, we can estimate certain features of the distribution of the mean from random sampling as follows: the sample mean $\bar{Y}$ estimates the mean of that distribution and $S/\sqrt{n}$ estimates its SD. Given a sufficiently large sample size, its shape will be well approximated by a Normal distribution, true for a wide variety of population shapes. This is a result of the CLT.

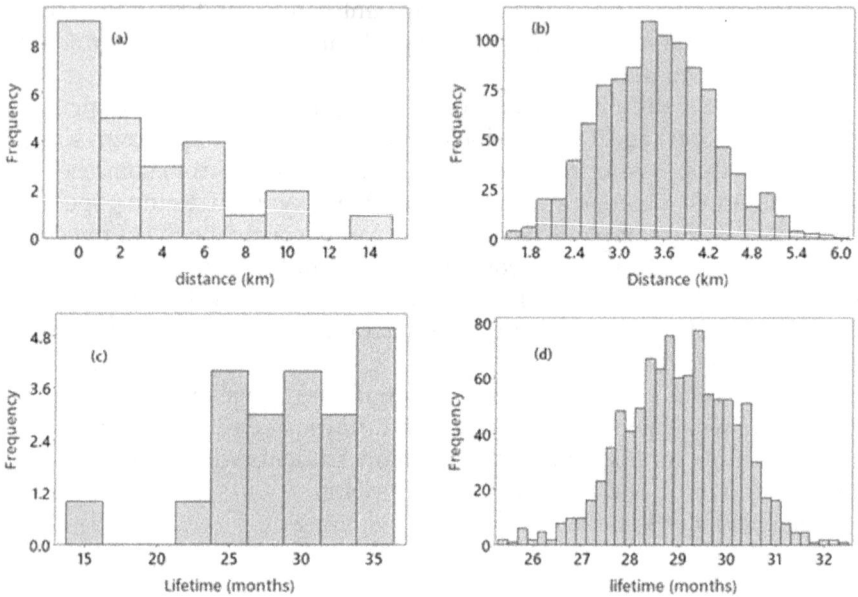

FIGURE 3.2
Two examples of skewed data (a and c) for which the distribution of the means (b and d) is approximately Normal.

How large a sample do you need for that result to obtain? Surprisingly, it is often not a very large number. We will illustrate that with a couple of examples.

The first one is a data set consisting of 25 measurements of distances between army ant colonies[7]. The data themselves are truncated at zero and sharply skewed right (Figure 3.2a). The distribution of the means (Figure 3.2b, from 10,000 simulations) is quite symmetric and bell-shaped. The second example consists of a sample of 21 lifespans from a mouse study. The data themselves are skewed left (Figure 3.2c), but the simulated distribution of the means is quite symmetric and bell-shaped (Figure 3.2d).

The basic premise behind the simple version of the CLT is that whenever you add two independent random variables, the distribution of the resulting variable will tend to be slightly closer to a Normal distribution than whatever you started with. We said, "add," but the notion generalizes to include subtraction and allows for multiplying variables by chosen constants. For example, $Y_1 + Y_2$, $Y_1 - Y_2$, and $2Y_1 + 3Y_2$ would all qualify; such formulations fall into a mathematical class called a "linear combination of the data values". The list of statistics given at the start of this chapter is all in that family of linear combinations. The classical statistical methods you learned to use in your first statistics classes all depend on Normality via the CLT. They do *not* depend on having data that are Normal. Rather, the question rests on whether the randomness in your data is "Normal enough."

3.2.1 Testing for Normality is Sometimes a Waste of Time

We tell students that testing for Normality is often a waste of time. This seems like a shocking thing for a statistician to say, but it does get their attention. There is enough truth to that statement, and it is sufficiently counter to common practice that we will show you why we say that. First, it is not the distribution of your sample that matters; what matters is the distribution of the population of values whence it came.

For small samples, you cannot trust the data to accurately reflect the population whence it came, so whether the sample is skewed or symmetric is not very relevant. This is illustrated in Figure 3.3, which shows four samples, each of size 7, from a distribution that is highly skewed and truncated at zero. We generated 20 such samples and chose illustrative examples from them. Only one (Figure 3.3d) was as distinctly skewed as the parent population. To be sure, a sample of size 7 is quite small, but it does illustrate that you cannot trust a small sample to reliably represent the population of interest.

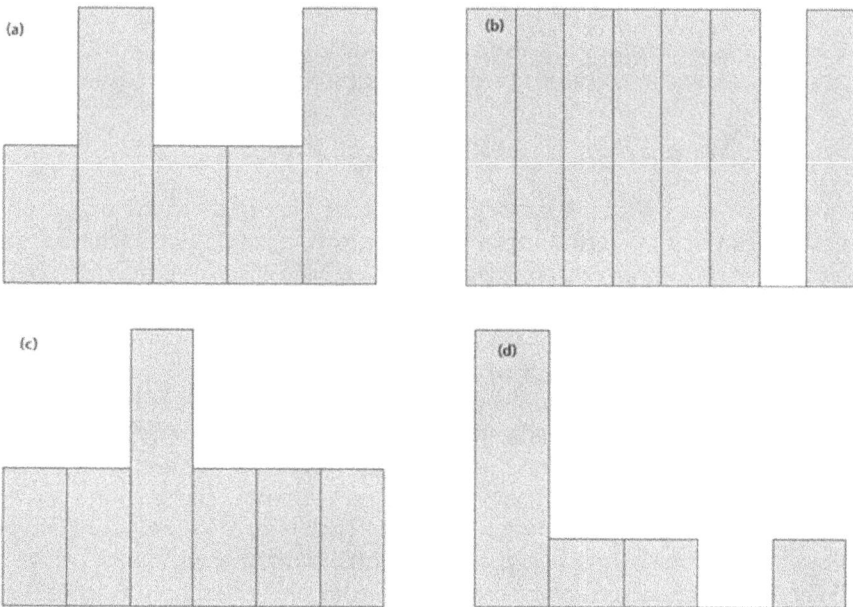

FIGURE 3.3
Four samples of size 7 from a highly skewed population.

For larger samples, the data will tell you more accurately about the distribution of the variable in the population, but you might not worry about it too much since the CLT might well be working for you. Good judgment is called for. The examples in Figure 3.2 illustrate this. In both cases, the sample histograms are skewed; furthermore, the shape is consistent with what you might expect from such data. A tropical ant ecologist would know that, in the right location, army ant colonies abound, and you won't go far from one until you bump into another. Occasionally, the inter-colony distance is large. Lifespans are an example of biological data that are commonly left-skewed. Most lab mice might live until 30 months or so, but an unfortunate few die young.

A test for Normality from a larger sample might well tell you that the data do not come from a Normal distribution, but as the simulations that generated Figure 3.2b and d illustrate, the CLT will deliver approximate Normality for the distribution of the means. If we consider larger and larger sample sizes, "Normal enough" can get even more relaxed.

Decades ago, we might have given you the advice that you need a sample of 30 or more to comfortably count on the CLT. We now know, thanks to high-speed computers that make simulations a snap, that this advice is conservative. Back in the day, it made sense because statisticians did not want you to use methods with shaky validity. Simply put, the more symmetric is the "parent distribution", the smaller is the sample size that will suffice. You cannot ever see the "parent distribution," of course, but you can simulate the process of repeating the experiment in order Bootstrapping (see Chapter 15 for details) is becoming more and more widely available in statistical packages. You can use it to assess whether invoking the CLT seems plausible.

3.2.2 Why Do We Discuss the Distribution of "Y" Anyway?

Oh, and one more thing. We discussed the CLT here in the context of studying the mean of a random sample. In that context, it is fair to ask whether the distribution of Y is Normal enough. If your analysis is more complicated than that, we discuss the distribution of the residuals from the model, not Y itself. Chapter 6, introducing simple linear regression, discusses this in more detail.

For analysis on a single mean from a random sample, there is indeed a model lurking behind the analysis, but it is so simple that it barely deserves the name: $Y = \mu + \varepsilon$. "Y equals some mean plus random error." Figure 3.4 shows waiting times between eruptions of a geyser and residuals from the relevant model: $res_i = Y_i - \bar{Y}$. The two histograms are not identical, due to software choices for placing numbers into bins. That said, a description of shape (bimodal, with the bigger bump on the right) would be the same.

When your data are composed of values from a single sample, it suffices to look at the distribution of the data themselves; calculating the residuals is an additional step that adds nothing to the evaluation. Indeed, for most of the 20th century, statistical calculations were done with very rudimentary tools (pencil and paper, anyone?). In those times, any computational shortcuts

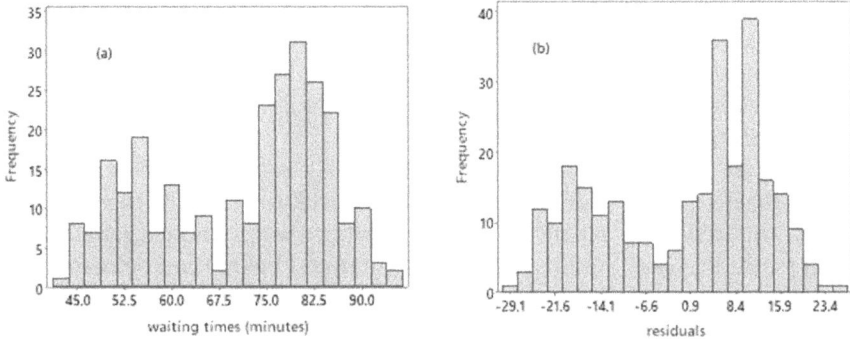

FIGURE 3.4
Histograms of waiting times and residuals (observed times minus sample mean) for the geyser data.

were quite appreciated. Most statistics classes start with single-sample analyses, for which you need only think about Y, and residuals might not get mentioned. And that, right there, is where the immortal, but flawed, question, "Are my data Normal?" arises along with its sequel declaration, "My data are skewed, so I log-transformed."

3.2.3 Life is Not Always Kind[8]

When we say that, in general, the distribution of a sample average will look like a Normal distribution, given a sufficient sample size, we mean it. But still, real life shows us not to take such things for granted. The data in Figure 3.5 illustrate this, for a sample of 96 mammalian body sizes in kilograms[9]. In that data set, the largest species is the African elephant; in second place is a hippopotamus.

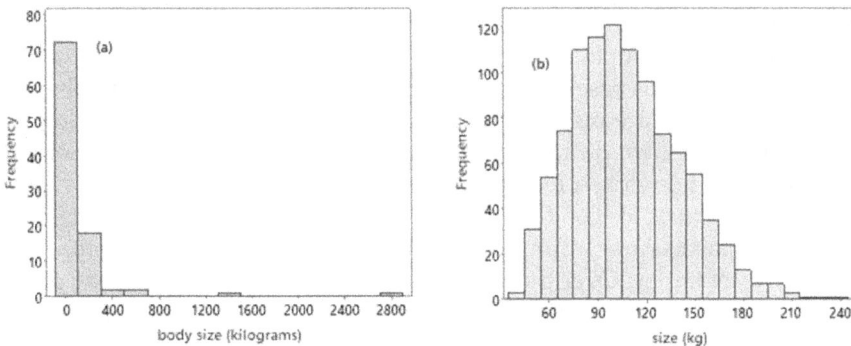

FIGURE 3.5
Adult body size for a sample of 96 mammal species. (a) The actual data; (b) a simulation of the distribution of the means from this sample.

Most mammals are quite small by comparison. The distribution of the data is so severely skewed that the distribution of the means is still clearly skewed.

Are there certain types of data that might impel us to feel uncertain? Data that are bounded, and when the mean is close to a boundary, are problematic. Here are some examples.

Count data (values cannot be less than zero) might be problematic when the average counts are low. As a quick test, check to see if the sample SD is large compared to the mean[10]. If so, then further examination (e.g. bootstrapping) might be wise. Sums of Likert data (defined below) can be reasonably treated as numerical, but there are lower and upper bounds to the values. If the mean is close to either boundary, be cautious.

You have all seen surveys where the requested response is on a numerical scale from 1 to 5, where 1 might indicate "strongly disagree" and 5 "strongly agree." These numbers are simply convenient labels for ordered categories. What moves one respondent to click on "5" might not move another. Another example you might be familiar with is a test with short-answer questions, each graded from 0 to 5. The grader might grade some questions more stiffly than others and might even grade differently right after lunch compared to right before lunch (graders are only human, after all). It is easy to see that these individual scores are not meaningful, in the fullest sense, as numbers. We routinely treat sums of such scores as though they were meaningfully numerical. That is likely fine except perhaps for situations where there are only a small number of responses summed together. If the mean is close to a boundary, the CLT might not work well for you.

3.3 Bait and Switch: We Mostly Don't Use the Normal Distribution Anyway

If you are familiar with doing analyses on statistically routine problems (e.g. paired data, two-sample data, regression, and ANOVA), you know that we usually use methods based on the *t*-distribution, not the Normal distribution. Exceptions to this are methods for Binomial proportions, including logistic regression. There, we do use the Normal distribution. William Gosset was a statistician who worked for Guinness Brewery in Dublin in the early years of the 20th century. He conducted many studies relating to beer quality, with an interest particularly in improving the yield and beer-making properties of varieties of barley. Over the years, he developed the sense that the distribution of a usual test statistic ($\frac{\bar{y}}{SE(\bar{y})}$) did not have a Normal distribution; rather, it was wider in the tails. Further, he even intuited that this feature diminished with larger sample sizes. To be sure, the two curves look similar (Figure 3.6). The phrase, "bell-shaped and symmetrical," could apply to each. That said, when you do statistical analyses (whether that be a test or a

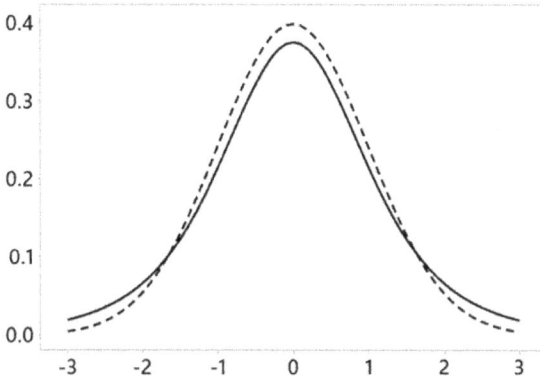

FIGURE 3.6
A standard[11] Normal distribution (dashed line) and Gosset's famous distribution (solid line), with four degrees of freedom.

confidence interval), the tails of the distribution matter. The endpoints of a 95% confidence interval are defined by values in the tails, and p-values from a test get interesting when they are small; small p-values are measurements from the tails of the distribution. We will study both intervals and tests in the next chapter.

He couldn't resist trying to figure out the formula. Armed with a pile of paper, some sharpened pencils, and a large eraser, he set to work. "Suppose Y comes from a Normal distribution..." It was straightforward (and well-known at the time) that the mean ($\bar{Y}$) also has a Normal distribution. Well, long story short, what Gosset landed on was[12]

$$\frac{\Gamma(n/2)}{\sqrt{(n-1)\pi}}\left(1+\frac{t^2}{n-1}\right)^{-n/2}.$$

Skipping details that are unimportant here, the important thing to note is that the formula now depends on the sample size n, sometimes appearing as n, sometimes as $n - 1$. He chose to call the $n - 1$ bit "degrees of freedom," as a convention arising from geometry, where it was the name associated with the dimensions of the geometric object under study.

Gosset's new distribution did indeed more accurately capture the behavior of his sample statistics (standardized means), and it thereby became a very useful tool in the arsenal of quality improvement tools for Guinness Brewery. It won't surprise you that his employers refused him permission to publish his new findings[13]. Eventually, they relented, but only on condition that he used a pseudonym so that the work could not be traced back to him (and hence, to Guinness). He chose, idiosyncratically, the name "Student" (Student, 1908) and called his distribution the t-distribution[14]. It revolutionized the practice of statistics, in particular, methods applied to small samples.

3.3.1 Nonparametrics aren't Necessarily the Solution Either

We have heard ever since the beginning of our careers that "our data were skewed, So..." followed by the assertion that the data were log-transformed or that nonparametric methods were used. We have devoted Chapter 9 to the role of logarithmic transformations, but we will briefly discuss nonparametric procedures here.

First, nonparametric procedures primarily focus on medians, not means. That's important because you should make clear at the outset which is of interest to you. If the population your data comes from is indeed skewed, then the population mean and median are meaningfully different. Means are often of more innate interest than medians, in no small part because differences in means directly imply differences in totals. See Section 2.3 for a fuller discussion of this. Second, nonparametric procedures don't automatically lend themselves to estimates (confidence intervals and such). You can get around this via, for instance, bootstrapping, but much of the time, users of nonparametric procedures report the test results and stop.

If your interest is indeed in making inferences about means, the CLT will often come to your rescue. If that appears not to be so, then randomization tests and bootstrapped confidence intervals may help you. If your interest is indeed in medians, then fine. Nonparametric procedures are what you need to use.

3.4 Chapter Summary

We introduced the notion of a distribution of a statistic as the answer to the question, "What would we imagine we would get if we repeated the study a very large number of times, and, every time, recorded the value of the mean?" We use the sample mean to estimate the mean of that distribution. The name "standard error" was chosen as the name for the standard deviation of the distribution of the statistic. So, as you will see in this text, the slope in a regression has its own SE formula, the difference in two means has its own, and so on.

We demonstrated that the place for the Normal distribution in classical statistics is in the distribution of your statistic, not in that of your data. This wonderfully convenient outcome (convenient because having data with a Normal distribution is not normal) is brought to you by the CLT. This result applies to statistics of random samples; random samples are not always easy to obtain (discussed in Chapter 2), and so invoking the CLT requires careful judgment.[15] For an example from this chapter, getting a random sample of waiting times between eruptions of Old Faithful geyser is implausible. Similarly, getting a random sample of distances between army and encounters is not feasible (you might strike out in a random direction, and then (reasonably) argue that the resulting data behave at least approximately like a random sample.

A relevant question for the application of the CLT to the distribution of a mean from a single sample[16] is whether the data are "Normal enough," with the word "enough" being a calibration word for sample size. The larger your sample size, the less you need to worry about (say) skewness in your data. We pointed out that, technically, the right question is whether residuals from your model are Normal enough, but for simple situations, it suffices to look at Y directly.

After spending the entire chapter arguing for the role of the Normal distribution, we summarily toss it out and use the t distribution instead. We told the story of William Gossett and how he was impelled to do the work of deriving this distribution. His discovery was a game-changer in the development of contemporary statistical methods.

3.5 Exercises

1. True or false and explain. Given random sampling, as the sample size increases, the distribution of the resulting samples will become more and more like a Normal distribution.

2. True or false and explain. Given random sampling, as the sample size increases, the distribution of the sample mean will become more and more like a Normal distribution.

3. With respect to the distribution of the mean from a random sample, what does the standard error (SE) of the mean estimate?

4. True or false and explain. As the sample size increases, the SE of a sample mean will tend to get smaller.

5. For each of the four scenarios below (Figure 3.7), making note of the shape of the distribution of the data, and the sample size, is the use of the t-distribution valid? Explain.

3.6 Solutions to Exercises

1. False. The sample will tend to look more and more like *whatever* the population of values looks like. More times than not, this will *not* be a Normal distribution.

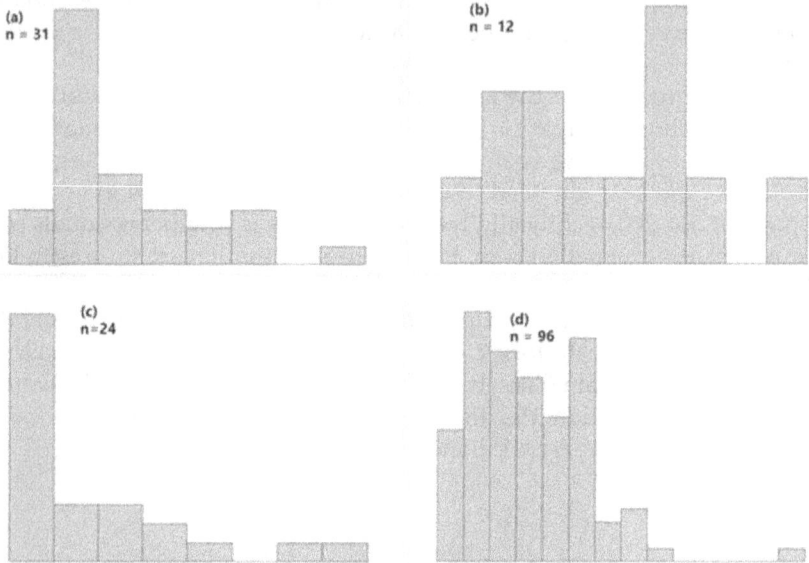

FIGURE 3.7
Four samples for Exercise 5.

2. True. This is precisely the gift that the CLT gives us.

3. The SE of the mean estimates the standard deviation of the distribution of the mean.

4. True. Intuitively, more data ought to yield better precision, which implies a smaller SE. Or you could note that sample size n is in the denominator of the SE formula.

5. In each case, we used bootstrapping to study the distribution of the mean via simulation. The results are collected in Figure 3.8, organized to match the layout of the question.

   a. The data are quite skewed, but the distribution of the mean is quite nicely symmetric and bell-shaped.

   b. The sample size is small, which makes one a bit nervous, but lo and behold: the distribution of the mean looks decently like a Normal distribution.

   c. Just to show that life is not always kind, in this case, the distribution of the mean has a distinct skew to the right.

   d. This distribution is approximately Normally distributed, but not perfectly so, by an old-fashioned ocular estimate. Its skewness is between that of examples (b) and (c)

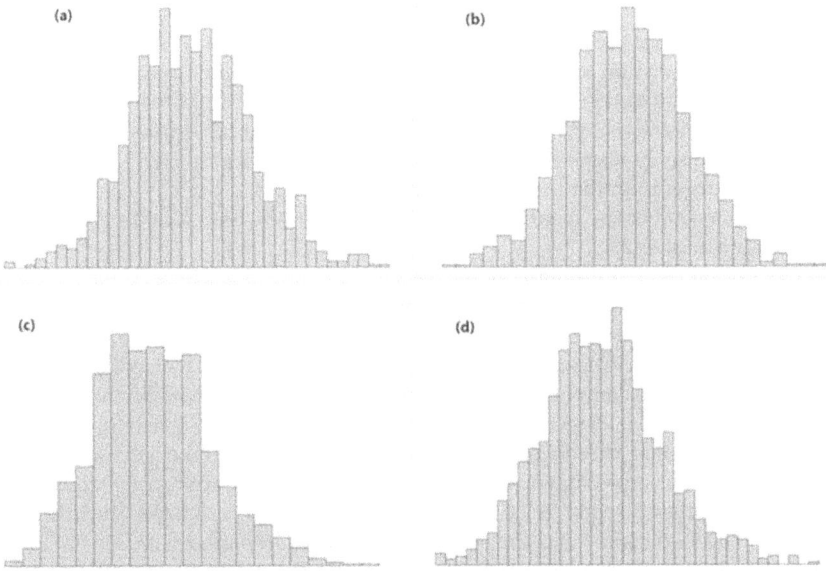

FIGURE 3.8
Distributions of the mean for Exercise 5.

Notes

1 Two things: most distributions of data seen in the wild are *not* Normal, but the CLT applies without concern for whatever distribution your data might have (we say, "most," and it applies very broadly). Also, the required sample size is not disturbingly large. In many cases, sample sizes in the teens will suffice.

2 If you keep track of proportion data with zeros for one category, and ones for the other, the sample proportion is the average of those (0, 1) values.

3 Never mind that the coin flip is now done, and so there is no longer any probability involved. *That* matter gets us off into realms of philosophy and existential angst that we would rather not consider here.

4 This is conventionally called the "sampling distribution of the mean." The adjective "sampling" is redundant, we think. We can just say "distribution of the mean" with no loss of meaning...

5 We will study these data in more detail in our introduction to simple linear regression.

6 You can think of this as a "thought experiment". No physical treatments or subjects are involved, only your mind and a computer to speed the arithmetic. We will study bootstrapping in more detail in Chapter 15.

7 Ken worked with these data at la Selva Research Station in Costa Rica in the 1990s. The source is lost in the mists of time.

8   This discussion is inspired and informed in part by discussions with Carson Keeter.

9   On average, for each species. These data are part of research studying gestation time as a function of body size. Gestation time is difficult to measure for some animals, due to their cryptic habits, but an estimate might help zoo or animal preserve managers protect pregnant females for a suitable amount of time.

10 If the SD is, say, about ½ the size of the mean or smaller, then you are likely fine. If it is larger than the mean, be wary.

11 A discussion of the meaning of "standard" is in the Statistics/English dictionary in Chapter 2.

12 We are leaving out details here like introducing the gamma function, and so on. If that opacity is troubling, you might need to take a statistics theory class to overcome those troubles.

13 A colleague had recently published some work and inadvertently gave away trade secrets to competitors. His employers were thus understandably nervous.

14 And so, we see it introduced in textbooks as "Student's t"... We tell our students he called it that after Britain's second favorite beverage (fake news).

15 And perhaps crossing your fingers or praying to some appropriate deity.

16 This applies directly to a few other statistics, for example, the difference between two means and the mean of differences in paired data.

References

Härdle, W. 1991. Smoothing Methods with Implementations in S. Springer Series in Statistics. Springer-Verlag, New York. 262 pages.

Student. 1908. The probable error of a mean. Biometrika 6: 1–25.

4

Using the Data: Introducing Hypothesis Tests and Confidence Intervals

Lab mice live for three or so years, depending on their circumstances. For our purposes, suppose the average lifespan for a certain strain of mice, fed on a normal protocol, is 31.3 months. A diet controlled at 85 kcal per week was fed to a sample of 57 mice. This is the usual average consumption, but in this case, the mice were held to it every week. The mean lifespan in the fixed diet sample was 32.7 months. Two questions come to mind:

1. 32.7 is indeed bigger than 31.3. But could that apparent increase in lifespan just be due to random chance? It is from a sample after all.

2. Related to that: our best estimate of lifespan for mice fed on a fixed 85 kcal diet is 32.7. The average in the population is most likely not this precise value. Can we place a margin of error (MoE) on this?

Every time you do a statistical analysis, you are doing one of two things: testing or estimating. This chapter is devoted to introducing those two big ideas.

A hypothesis test begins with a research hypothesis or question. This may have a sense of direction: is there a positive relationship between two variables? Is the average of this group larger than that of another group? or not: is there a relationship? Are the averages different? Then that hypothesis gets nullified, and the data are examined to see if there is sufficient evidence with which to reject the resulting null hypothesis. In broad brushstrokes, the process is very much like a criminal court case. Instead of seeking evidence "beyond a reasonable doubt," in a hypothesis test, you choose the amount of risk you are willing to take of rejecting the null when in fact it is true. This is the so-called alpha-level, often chosen to be 0.05. We will cover details of how to do testing, including one- and two-tailed tests.

Numerical estimates benefit from some statement of precision. At the very least, many researchers present the standard error of the estimate. A confidence interval takes that statement to its fullest expression. We will show you how confidence intervals behave depending on the choice of confidence level, sample size, and innate variation in the data and explain why they are called "confidence intervals." We will then demonstrate the construction of some standard intervals.

DOI: 10.1201/9781003609605-5

4.1 Introduction to Hypothesis Testing

In this section, we will introduce you

- Construction of appropriate hypotheses,
- The idea and use of a *p*-value as evidence in a hypothesis test, and
- How and when to use one-tailed versus two-tailed tests.

Full disclosure: this lifespan example, based on real data, is somewhat contrived. We owe you an explanation. Almost never will you do research with only a single sample. Almost all studies are more complicated. We chose to introduce hypothesis testing in this single-sample setting precisely because it is the simplest, and it allows us to focus on testing concepts and details, without any distractions. If you do a test on the mean from a single sample, you must have some value and a reason for its existence from outside your data. There is no obvious choice "baked in" to the situation. The contrivance? We made up that value of 31.3 weeks. It was chosen because we hope it leads to helpful discussion here.

4.1.1 Null and Alternate Hypotheses, Alpha

A good analogy to scientific hypothesis testing is a criminal trial. An individual is brought to trial if there is some reason to believe she/he may have committed a particular crime. The starting point for the trial is a presumption of innocence *of that crime*, followed by consideration of whether the evidence is consistent with that innocence. First, you name the crime, then presume it didn't happen, and see if the facts of the matter line up with that presumption.

You might recall being given a scenario in an introductory statistics class and asked to write down the relevant hypotheses, then follow through with the test. So you look in your notes or text for guidance. Aha! Borrowing numbers from our example:

$$H_0 : \mu = 31.3$$
$$H_A : \mu \neq 31.3$$

where H_0 symbolizes the null hypothesis and H_A is the alternate. After some head scratching or maybe even struggling, you put it together, and on you went. Still, you might recall that it was not easy. There is a reason for that. **It is backwards!** True, this is the conventional way to present the hypotheses, but let us say it again: it is backward. A criminal court case does not start with the presumption of innocence[1]. It starts with the presumption of innocence *of the crime* in question. You have to say, innocent *of what*. Without the" what," there is no trial. And that, in a nutshell, is the problem with the usual way to present hypotheses. You must first specify the alternative and then nullify it. As you'll see below, the specification of the alternate matters. Let's start by writing them in a smarter order:

$$H_A : \mu \neq 31.3$$
$$H_0 : \mu = 31.3$$

And while we are at it, it's not just some alternative hypothesis, it's your research hypothesis! So better yet would be

$$H_R : \mu \neq 31.3$$
$$H_0 : \mu = 31.3$$

Moving forward, you will probably write them in the usual order with usual labels, since doing so is such a strong convention. Here we are going to write them in our preferred order. Maybe we will start a revolution.

The process starts with a question, sometimes an actual hypothesis. Here it is, "Does regulating the diet affect lifespan?" The question then needs to be sharpened to take aim at a parameter of the population; here that would be the mean, leading to $H_R : \mu \neq 31.3$, which is a crisp and symbolic way to express the research question as a statement about the mean lifespan in the population. Two variations to this question could have been posed. Here are all three scenarios, along with the hypotheses they line up with.

1. Does regulating the diet affect lifespan? $H_R : \mu \neq 31.3$.
2. Does regulating diet increase lifespan? $H_R : \mu > 31.3$.
3. Does regulating the diet decrease lifespan? $H_R : \mu < 31.3$.

And here are all three, along with their nullified counterparts:

$$H_R : \mu \neq 31.3 \quad H_R : \mu > 31.3 \quad H_R : \mu < 31.3$$
$$H_0 : \mu = 31.3 \quad H_0 : \mu \leq 31.3 \quad H_0 : \mu \geq 31.3$$

In all three cases, you will use 31.3 as your test value. Once we pin down how to do the test, we will return to these three for a moment of "compare and contrast." We note here that scenario (1) leads to a so-called "two-tailed" test, while the other two are "one-tailed".

In a court trial, there is some chance that an innocent person is found guilty. In a hypothesis test, there is a chance that we could reject the null when in fact it is true, a so-called false rejection or false significance. In the court case, the jury is told to seek "evidence beyond a reasonable doubt.[2]" In a hypothesis test, we *choose* the amount of risk we are willing to take of a false significance. Conventionally, this chance, the so-called alpha-level, is set at $\alpha = 0.05$. It reflects the sense among scientists that in most situations, 1 chance in 20 of a false significance is an acceptable risk. Larger values (0.10, 0.15, say) are often used in pilot studies, where the consequence of false significance is smaller. Lower alpha levels (0.01 or 0.005, say) might be used when a false significance has large and unpalatable consequences.

4.1.2 Assessing the Evidence in a Hypothesis Test: The *p*-Value

The test statistic here is calculated as $t = \frac{\bar{y}-31.3}{SE(\bar{y})} = \frac{32.7-31.3}{0.68} = 2.05$, reporting that our observed mean is 2.05 SEs above the hypothesized value. The distribution used to do the test is a t distribution, with a mean of zero, which is consistent with the null that the true mean lifespan is 31.3, and with specified degrees of freedom. Here $n-1 = 57-1 = 56$. Simply put, the further the observed value is from the test value, the more evidence there is that the observed difference is not merely due to chance.[3]

The *p*-value for a hypothesis test is the probability of obtaining a test statistic as extreme or more extreme than the one observed, assuming the null hypothesis to be correct. This is a mouthful of words and ideas. More simply, for your observed statistic, determine how much of the relevant t distribution is even *further* out in the tails. If that value (the *p*-value) is small, your test statistic is far into the tails and hence far from the null. To interpret a *p*-value, compare it to alpha. If the *p*-value is less than alpha, reject the null hypothesis.

Let's do the tests for our three example scenarios.

> **Scenario 1.** $H_R : \mu \neq 31.3; H_0 : \mu = 31.3$. In this case, if the null hypothesis is not true, the true value could equally well be less than or greater than. Accordingly, the *p*-value calculation takes both tails of the distribution into account. Here (Figure 4.1a), $p = 0.045$. This is slightly less than $\alpha = 0.05$, so we would say we have modest evidence that the diet increased, average lifespans.
>
> **Scenario 2.** $H_R : \mu > 31.3; H_0 : \mu \leq 31.3$. Here, we are only interested in asking if the mean lifespan is larger than 31.3; smaller values are consistent with the null. Accordingly, the *p*-value is 0.0225 (Figure 4.1b), and we could claim to have good evidence that the regulated diet caused an increase in mean life expectancy.
>
> **Scenario 3.** $H_R : \mu < 31.3; H_0 : \mu \geq 31.3$. The research question was whether there had been a *decrease* in lifespan. Our observed mean is greater than 31.3, so the test is over. Go grab coffee or lunch and try something else this afternoon. We guarantee that the *p*-value will be greater than 0.5, which you can see play out in Figure 4.1c (*p*-value = 0.9775).

In a statistical hypothesis test, when the data are not consistent with the null hypothesis, we say that we reject the null. Another term of art among scientists is that the result is "statistically significant," meaning only that the effect appears to be real.[4,5] Statistically significant does *not* imply "significant" or "important" as it would in everyday English. This interpretation is

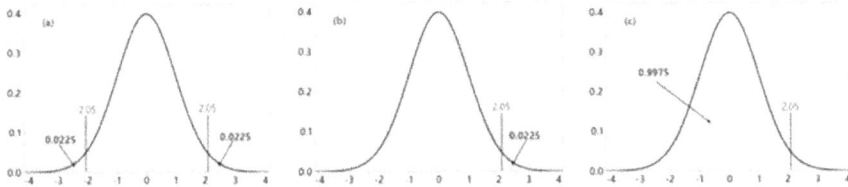

FIGURE 4.1
Illustration of *p*-value calculations for three scenarios.

so incredibly seductive that the American Statistical Association made the foregoing point in a formal editorial statement (Wasserstein and Lazore, 2016, cited in Little, 2025, page 37).

In fact, Ron Wasserstein and Nicole Lazar felt that the statement did not go far enough and proposed that we declare "statistically significant" to be *persona non grata*: "In sum, 'statistically significant'—don't say it and don't use it." (Wasserstein et al, 2019, page 2) and declared that by, "breaking free from the bonds of statistical significance, statistics in science and policy will become more significant than ever." (page 10).

The *p*-value approach allows flexibility for readers of test results. For example, suppose someone reports $p = 0.07$ in a comparison of two groups, and declares the result not significant, against the conventional $\alpha = 0.05$. And suppose further that you are reading the paper for ideas for inclusion in a research project of your own. You are completely free to gauge their *p*-value against, say $\alpha = 0.15$, which might be a reasonable choice for use in a pilot or preliminary study, and declare, for your purposes, that the result is indeed significant. Or, on the other hand, suppose someone reports a *p*-value of 0.03, which they declare significant against the conventional alpha. If you are considering investing a large amount of time or money into the studied treatment, you might have $\alpha = 0.005$ in mind since you are unwilling to risk your resources on something without solid evidence, and so, for your purposes, find the result to be not significant.

4.1.3 Perfection, If You Want It

As we have shown, you are free to choose alpha; never mind that most of you will most of the time use $\alpha = 0.05$, since it is such a strong cultural norm. If you use a larger alpha, you increase your chances of finding a "significant" result at the price of increased risk of "false significance"; a smaller alpha gives you more protection against the latter. You pay for that with reduced chances of a "significant effect" finding. It is up to you. For example, if you want to claim a new variety of corn has a higher yield than those currently in use, you better be very sure, or farmers may blame you for loss of income. In that case, you might choose a smaller alpha.

Suppose you really are worried about missing interesting findings. In fact, you lie awake at night worrying about that. Navarro's testing method will *guarantee* that you never miss an interesting finding ever again.

> **Navarro's testing method (guaranteed to not miss an effect):** Choose $\alpha = 1$. Any *p*-value you calculate from real data is guaranteed to be less than that. Statistical significance every time!
> On the other hand, suppose you really are worried about the embarrassment of declaring in favor of a significant result when in fact there is no effect. In fact, you lie awake at night worrying about that. Gerow's testing method will guarantee that you will never suffer that terrible fate.

> **Gerow's testing method (100% guaranteed to protect against a false significance):** Choose $\alpha = 0$. You will never get a *p*-value less than that! A particular benefit of either of these methods is that you don't even have to analyze your data. In fact, you don't even need to go to the bother of collecting any data[6].

4.1.4 Thoughts on One- versus Two-Tailed Tests

If you compare panels (a) and (b) of Figure 4.1, you see the advantage of having a "sense of direction" to your research question. If indeed the data point is in the same direction as your hypothesis, the resulting *p*-value will be ½ of the analogous two-tailed test. This, in turn, implies that the test has more power. If the effect is consistent with your hypothesized direction, you are more likely to declare in favor of it. This is the meaning of power in the context of a statistical test. While it is true that a one-tailed test is more powerful than the analogous two-tailed test, there are some cautionary points worth mentioning.

First, in the conventional testing paradigm, the decision to declare a sense of direction to the test must come before the data are collected, as part of the study design stage. By the time you have collected the data, the direction of any effect is likely to have come to your attention, even if only subliminally.

There are two common bases for choosing a one-tailed test. One is theory or historical evidence. For instance, adult male grizzly and black bears are, on average, larger than their female counterparts. Suppose a new bear species is discovered in the remote Lost Land Archipelago.[7] You and your colleagues anticipate capturing a number of them to collect biometric data, including weight. It would be quite reasonable to hypothesize that the males might, on average, outweigh the females. This basis runs the risk that the pattern might not play out in that Lost Land species. The other basis is simply interest. If we come up with a new treatment that we hope will shorten the duration of the common cold, our interest is only in one direction: on average, the mean duration is shorter in the treated group as compared to the control group.

If the data points in the wrong direction (Figure 4.1c), your directional hypothesis is toast. Back to the drawing board. People don't do one-tailed tests as often as you might think, and in truth, it often doesn't make much difference in scientific work. If the two-tailed p-value is large (0.3, say), the one-tailed analogue (0.15) would still lead to a declaration of "no significant effect." If the two-tailed p-value is very small (0.001, say), then cutting it in half would not particularly change your conclusions. But let's say the two-tailed p-value is 0.06. Against the conventional $\alpha = 0.05$, this is not small enough to lead to a rejection of the null. The one-tailed value would be 0.03, which would lead to a statement of a "statistically significant finding." Not by much, to be sure...

In some situations, the binary "reject, fail to reject" nature of a hypothesis test leads to crisp decisions. For example, a truckload of computer chips arrives at your factory, and you test a sample of them, looking for defects. Based on the results, you either unload the truck or send the whole load back whence it came. A binary outcome. Or, say, in a wildlife management setting, you might have to decide on whether to allow hunting in a given season. You either do or don't, and so a test of abundance from a sample has real and immediate consequences.

In scientific research, however, we don't often have to make a crisp decision based on test outcomes. In fact, it is usual when reporting p-values from several tests for scientists to simply express their feelings about the p-values by adding more asterisks to express their excitement. For example, "* for $p < 0.05$, ** for $p < 0.01$, and *** for $p < 0.001$". A p-value is in fact a statistic (a summary number arising from your data), and so if you imagine repeating your study repeatedly, you should expect some variation in the p-values. In that context, $p = 0.06$ and $p = 0.03$ are not that different from each other.

4.2 Confidence Intervals

In scientific practice, it does not suffice to simply give the numeric value of an estimate. For example, stating that the average lifespan in the sample of 57 mice on the fixed diet was 32.7 weeks would leave your reader wanting more. How close do you think your estimate is to the population mean μ? A usual way to answer this question is with a confidence interval (CI). For the mouse data, a 95% CI is given as (31.3, 34.1). The interpretation is, "we are 95% sure that μ is between 31.3 and 34.1." A little later, we will look under the hood to see how it is calculated, but for now, consider:

- Why 95%
- What would change if we chose 90% or 99%?
- Why is it called a 95% interval?

The choice of 95% as a confidence level is a very strong cultural norm, but there is a rationale behind it. As we discussed above, $\alpha = 0.05$ is the conventional choice of alpha-level for a hypothesis test. It makes sense to choose your confidence level to be complementary to your alpha level.[8] For instance, $\alpha = 0.10$ ought to be accompanied by the choice of 90% as a confidence level.

The other questions focus on the properties of CIs, so let's start there.

4.2.1 Properties of Confidence Intervals: How Do They Behave?

Figure 4.2 shows 20 CIs generated from random samples for combinations of sample sizes 10 (top row) and 30 (bottom row) and confidence levels 90% (left column) and 99% (right).

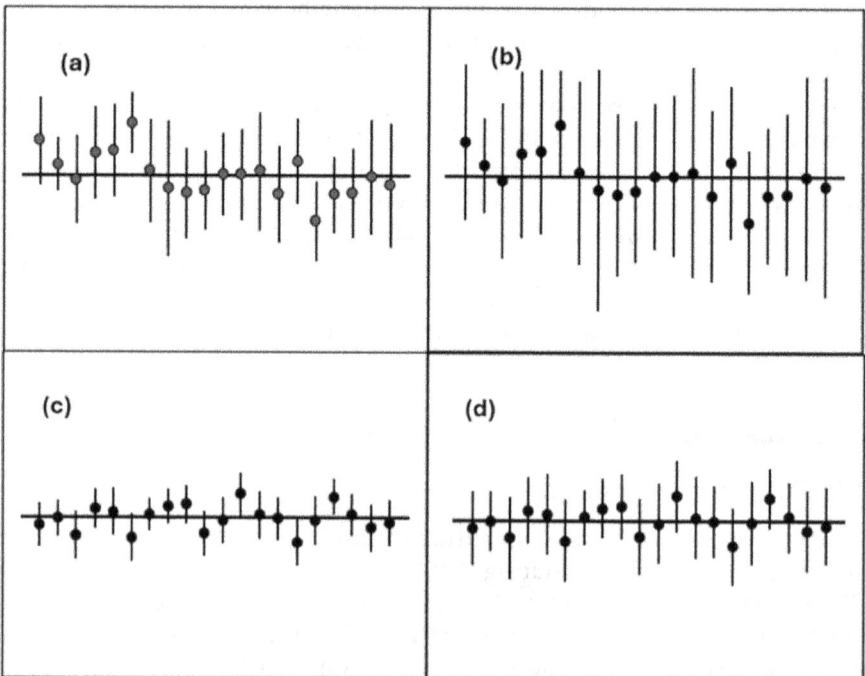

FIGURE 4.2

Depiction of 20 confidence intervals (vertical lines) from a Normal distribution for each of four scenarios. For panels (a) and (b), the sample size was set at $n = 10$; for (c) and (d), it was $n = 30$. In panels (a) and (c), 90% is the confidence level; in panels (b) and (d), it is 99%.

Notes: the horizontal lines depict μ, while the black circles show each sample mean. The vertical lines emanating from the sample means show the range of each CI.

Take-home points from studying Figure 4.2:

1. In each panel, the sample means vary from one to another, and so do the interval lengths. This is to be expected: if you were to replicate an experiment a second time, you would randomly get different data. Hence, different means and SDs (which determine the length of the intervals).

2. In panel (a), 2 of the 20 intervals do not contain μ. In panel (c), 3 do not. In panels (b) and (d), none do (one *almost* does in panel b). This speaks to the following technical property of CIs. Given your choice of confidence level (90% for the sake of example), if you were to repeatedly repeat your study with such intervals, then 90% of them would contain the parameter being estimated. Those percentages don't show up perfectly when you only have 20 trials.

3. Looking from panel (a) to (b) (90% to 99%), you can see that the intervals are all wider. Ditto from panel (c) to (d). In fact, the same samples were used in (a) and (b), and a different set of the same samples in (c) and (d). Our point? With a given amount of information, if you need to be more sure of your conclusion, you need to use wider intervals. There is a price to that surety.

4. Now look from panels (a) to (c) and then (b) to (d). The samples in the bottom row are samples of size 30; the top ones are of size 10. What do you get for the price of collecting more data? The intervals are all narrower. That's a good thing.

5. Increasing the sample size will purchase smaller intervals, yes. But it does *not* buy you more surety. If you declare you want to be 90% confident, then the procedure will lead to intervals that capture μ 90% of the time. Ditto for 99% confidence level. Notice that the intervals in (d) are approximately the same length as the intervals in (a). If you want to be more sure (99% as compared to 90%) *and* you want narrow intervals, you will have to pay with a larger sample size.

Now for a brief introduction to the construction of standard intervals.

4.2.2 Construction of a Standard Confidence Interval

We will study the case of a CI for a mean, so you can better understand the general principles behind their construction. Where an explicit confidence level is helpful, we will use 95%.

The formula for a 95% CI for the mean is $\bar{Y} \pm t_{n-1,0.95} \times SE(\bar{Y})$, where the "±" part of the formula is sometimes called the MoE.

The logic behind this formula hinges on the Central Limit Theorem and William Gosset's discovery of the t distribution (Chapter 3).

For those who want to look under the hood, we will now show you some of those computational details. We won't continue with these details throughout the book, but you will see from these examples that the general construction principles are the same; details, standard errors in particular, differ.

Example 1. The lifespan for 57 mice on the fixed calorie diet was 32.7 months; the sample SD was 5.125 months. What is a 95% CI for the true average lifespan of mice like these? The standard error of a mean from a simple random sample is estimated by $SE(\bar{y}) = \frac{s}{\sqrt{n}} = \frac{5.125}{\sqrt{57}} = 0.68$. Here, $t_{56,0.95} = 2.00$. Therefore, the MoE is $2 \times 0.68 = 1.36$. We are 95% sure the true mean lifespan is between $32.7 - 1.36 = 31.34$ months and $32.7 + 1.36 = 34.06$ months.

Example 2. In addition to the foregoing data, suppose the mean lifespan for 71 mice on a reduced-calorie diet was 42.3 months; the sample SD was 7.77 months. Calculate a 95% CI for the true difference in mean lifespans of mice on diets like these. The parameter of interest is the difference[9] $\mu_1 - \mu_2$; the sample difference is $(\bar{y}_2 - \bar{y}_1) = 9.6$. The SE of the difference in two independent means from simple random samples is estimated by $SE(\bar{y}_2 - \bar{y}_1) = \sqrt{\frac{s_1^2}{n_1} + \frac{s_2^2}{n_2}} = \sqrt{\frac{5.125^2}{57} + \frac{7.77^2}{71}} = 1.145$. The MoE for a standard interval for the mean is computed as $t_{df,CL} \times SE(\bar{y}_2 - \bar{y}_1)$. For now, we'll just tell you that $df = 121$ for this data set. With $t_{121,0.95} = 1.98$ and the MoE is $1.98 \times 1.145 = 2.27$. Hence, we are 95% confident that the true difference in mean lifespans is between $9.6 - 2.27 = 7.33$ months and $9.6 + 2.27 = 11.87$ months.

4.3 Chapter Summary

We showed you that the usual order of presentation of hypotheses, first the null, then the alternative, is inconsistent with how they must be created. Do as you wish. The contemporary way to do a hypothesis test uses a *p*-value and is simple to deploy: if *p* is less than alpha, reject the null; otherwise, do not. This approach is flexible to use: a reader can compare the *p*-value from a study to their own choice of alpha and legitimately arrive at their own conclusion.

If your research question has a sense of direction, you can deploy a one-tailed test. If the observed effect is consistent with the direction of your hypothesis, then the resulting *p*-value will be ½ that of the analogous two-tailed test. Thus, when appropriate, a one-tailed test is more powerful than a two-tailed test.

Just as it is conventional to use $\alpha = 0.05$, it is usual to choose 95% as the confidence level when making CIs. These two values are said to be complementary to each other.[10] For another example, 90% is the complementary confidence level to $\alpha = 0.10$. We learned that a 99% CI will be wider than a 90%.

Simply put, if you need to be more sure that you have captured the parameter of interest, you need a wider interval. CI width goes down with increasing sample size; this is intuitively appealing, since more data ought to give you better precision (i.e. a smaller SE).

For some simple cases (one and two sample settings for means, we worked through CI construction. Certain details (SE formulas in particular) differ from one situation to another, but the basics stay the same.

4.4 Exercises

1. In a certain situation, two individuals calculated CIs for the mean from the same data. One interval was (10.5, 19.5), while the other was (12, 18). One used 90% as their confidence level; the other used 95%. Which interval was created using 90%? Explain.

2. True or false and explain. When doing a hypothesis test, if the resulting p-value is greater than α, you reject the null hypothesis and declare a significant effect.

3. In a certain situation, a researcher hypothesized that a novel treatment would increase the response variable over values in a control treatment: her research hypothesis was $\mu_T - \mu_C > 0$. The observed difference was $\bar{y}_T - \bar{y}_C = -2.3$. What was the resulting p-value from this test?

4. In a study of nest success, a biologist counted the number of fledglings in a random sample of 40 nests. "Since the researcher has only a single sample, not two samples, a one-tailed hypothesis test is called for." Is that statement correct or incorrect? Explain.

5. A biologist studying ferruginous hawk (FH) nesting behavior did a survey of 40 ten-square-mile study plots, counting the number of FH fledglings found in each plot. Historically, there have been, on average, 2.4 fledglings per ten square miles in this region. Have management interventions designed to minimize human impact during nesting season led to an increase in the average number of fledglings? Which of the following computer outputs is the correct one for this situation?

(a) Test of mu = 2.4 versus > 2.4				
n	Mean	SD	SE Mean	p
40	2.90	1.70	0.27	0.035
(b) Test of mu = 2.4 versus not = 2.4				
n	Mean	SD	SE Mean	p
40	2.90	1.70	0.27	0.070

(c) Test of mu = 2.4 versus < 2.4				
n	Mean	SD	SE Mean	p
40	2.90	1.70	0.27	0.965

6. State conclusions from the test in question 4, in terms of the research problem; i.e. take it beyond "reject/fail to reject" language.

7. In media reports of polling results, you often see conclusions along the lines of, "56% of voters say they will vote for Tyler, with a MoE of 2.5%, accurate nineteen times in twenty." Where do the margins of error and accuracy conclusions originate?

4.5 Solutions to Exercises

1. To be more sure of having captured the parameter in your interval (i.e. have a higher level of confidence), you need a wider interval. If you are willing to be less sure, your interval can be narrower. So, the (12, 18) interval was created using 90% confidence.

2. This is false. Reject if p is less than alpha. While to some, the definition of the p-value is arcane, its use is very simple.

3. You would have to have the data in hand and a statistics package to get the actual value, but we can be 100% certain that it is greater than 0.5, since the observed effect is running in the opposite direction to the research hypothesis.

4. This is not correct. Whether a test is one-tailed or two-tailed depends on whether or not the research hypothesis has "a sense of direction." If it does, the test is one-tailed. If not, then it is two-tailed. This is without regard to whether the setting is a one-sample, two-sample, or paired samples setting. For a single sample, in order to do a test for the mean, you have to have some "target value" in mind. In this current case, for instance, you might know the historical average and want to test to see if that is still the case.

5. The correct choice is (a). It is the output for a one-tailed test seeking evidence of an increase.

6. Output (a) is the correct answer to the preceding question, but here are the conclusions from all three, so that your answer here will be consistent with your answer from question five and will keep the conclusions consistent with the research question associated with each case. If you chose (a), "We conclude that there has been an increase in the average number of fledglings. For (b), you would conclude that

there is only weak evidence, not quite statistically significant, of a change in the average number of fledglings. For (c), "There is clearly no evidence that fledgling abundance as decreased."

7. The reporters created a 95% CI (hence the 19 times out of 20), the MoE of which was 2.5%. The entire interval, had they reported it, is (53.5%, 58.5%). They chose the language to represent the statistical procedure in a manner that would make intuitive sense to a reader, without forcing the reader to have taken a statistics class. Some might consider this an act of kindness.

Notes

1 It's not possible! None of us have been innocent since about age three, when we told our parents our first little white lie.
2 That instruction is deliberately vague because, for instance, the consequences of a mistaken "guilty" finding are very different for petty theft than for murder. Presumably, the jury would demand stronger evidence for the latter.
3 For tests with a sense of direction (tests 2 and 3), we must pay attention to direction; we will come back to that.
4 We use the word "effect" here as a generic term. In practice, it might get replaced by "difference in two means (or proportions)" or "the slope in a regression model."
5 In many instances, the null value (for an effect) is zero, so here the word, "real" simply says, "not zero."
6 We suggest that if you used one of these methods and share the results with collaborators or your supervisor, doing so on April 1 would be auspicious.
7 Don't bother looking it up. Fake news…
8 The reasoning, along with examples, is in Chapter 5.
9 We choose to subtract the smaller mean from the larger one, to make the story easier to tell.
10 Choosing them to be complementary is a good idea, which we will discuss in Chapter 5.

References

Little, R. 2025. Seminal Ideas and Controversies in Statistics. Chapman and Hall, 223 pages.

Wasserstein, R.L. and N.A. Lazar. 2016. The ASA's statement on p-values: context, process, and purpose. American Statistician 70(2): 129–133.

Wasserstein, R.L., A.L. Schirm, and N.A. Lazar. 2019. Moving to a world beyond p < 0.05. American Statistician 73: 1–19.

5

How Confidence Intervals and Tests Play Well Together (or Not)

Just as $\alpha = 0.05$ is a conventional choice of significance level for hypothesis tests, so is 95% a conventional confidence level for confidence intervals (CIs). Not everyone reports CIs; for sake of saving space, many journal articles simply report the estimate of interest and its standard error. We show you that you can quickly calculate an approximate 95% CI using the number "2".

A confidence level of 95% and $\alpha = 0.05$ are considered complementary values. For another example, $\alpha = 0.10$ is complementary to a 90% confidence level. Cultural conventions being what they are, most folks use the 0.05, 95% pairing without a pause to reflect on the fact that the two are each their own choice. However, we will show you why choosing them to be complementary is a good idea.

Sometimes researchers present several graphical summaries of means along with lines that indicate margins of error for the CIs. It is tempting to examine these and make tentative conclusions of statistical significance based on whether the intervals overlap (not significant) or do not (the differences are significant). Reality is a little more subtle; we will examine and discuss.

It is becoming common in statistics packages that if you ask for a one-tailed test, the package will deliver a so-called "one-sided confidence interval," also known as a (lower or upper) confidence bound. We discuss the rationale for that and also the issues that may arise.

And then, because we cannot help ourselves, we will give you our soapbox take on the relative utility of tests and estimates. This discussion might or might not be slightly controversial to you.

5.1 Confidence Level of 95% and the Number "2"

The choice of 95% for a confidence level is ubiquitous. Most statistics you use will fall into the list for which the Central Limit Theorem applies, namely means, differences in means, Binomial proportions, regression coefficients, and a few others. Historically, researchers used statistics from this list almost exclusively, precisely because those statistics were the only ones statisticians could easily provide guidance for. That is no longer the case. Statistics packages routinely produce CIs for medians (via an algorithm, not an equation),

DOI: 10.1201/9781003609605-6

TABLE 5.1

The *t*-multipliers for 99%, 95%, and 80% Confidence Intervals for the Mean of a Single Sample from a Variety of Sample Sizes *n* (Recall that Degrees of Freedom are *n*-1)

Confidence Level	*n* = 5	*n* = 10	*n* = 15	*n* = 20	*n* = 30
99%	$t_{4,99} = 4.60$	$t_{9,99} = 3.25$	$t_{14,99} = 2.98$	$t_{19,99} = 2.86$	$t_{29,99} = 2.76$
95%	$t_{4,95} = 2.78$	$t_{9,95} = 2.26$	$t_{14,95} = 2.14$	$t_{19,95} = 2.09$	$t_{29,95} = 2.05$
80%	$t_{4,80} = 1.53$	$t_{9,80} = 1.38$	$t_{14,80} = 1.35$	$t_{19,80} = 1.33$	$t_{29,80} = 1.31$

for instance. And bootstrapping can be used in almost any situation. We introduce bootstrapping in Chapter 15, but for a deeper introduction, see Chapter 2 of Manly and Navarro Alberto (2021).

In the case of the classical tools, it is worth pointing out that for all but small sample sizes, the multiplier in the margin of error for a 95% CI is very close to 2. For the sake of illustration, Table 5.1 shows the *t*-multiplier for a range of both confidence levels and sample sizes.

It is common for a written report to show the standard error of a statistic along with the estimate. This is more compact than showing the CI itself. In that case, it would not be very wrong to calculate your own interval using "2." You might, for your own safety, call it an approximate interval. Indeed, Scheaffer *et al* (2012), in their sampling methods text, use the number 2, and call "$2 \times SE$" a bound on the error. Quite simplified, and not unreasonable. If you are feeling sassy, you could call it a PDS interval[1].

5.2 Complementary Confidence Level and Alpha: It's Not the Law, but is a Good Idea...

We will illustrate the point using the mouse lifespan data introduced in Chapter 4. The two-tailed test against the null value of 31.3 yielded a *p*-value of 0.045, a marginally significant effect when using $\alpha = 0.05$. The complementary 95% CI is (31.33, 34.05). The *p*-value squeaks in under 0.05, and the interval barely does not include the null value. In that sense, the test and estimate tell the same story. If you choose a higher confidence level, the resulting interval will be wider. Indeed, a 99% CI is (30.9, 34.5). So now the test barely rejects 31.3, but the interval suggests that 31.3 is barely in the list of possible values. The two sides of the storytelling appear to be in conflict.

The opposite mismatch can also occur. Suppose that in the mouse study, the *p*-value had been 0.06: the findings are not statistically significant (barely). In that case, a 90% CI (narrower than the analogous 95% interval) would exclude the null. Once again, the two sides of the analysis don't quite tell the same tale.

There is no need to get all wound up about this problem, because we are 95% confident that you will default to use $\alpha = 0.05$, and choose 95% as the confidence level, just because that's what we all do most of the time. It is also the case that the mismatch will be obvious only when the *p*-value from the test is quite close to alpha. As a generally good idea, we recommend matching the confidence level to the alpha level so as to maintain consistency in the findings from each side (test, interval) of the analysis. We will have more to say about intervals and tests in Section 5.5.

5.3 Do Overlapping Confidence Intervals Coincide with a Non-Significant Test?

When looking at CIs for means from two independent samples, it is tempting to infer test results based on whether the intervals overlap. If they do not overlap, they declare that there must be a significant difference between the two population means[2]; if they do overlap, there is none. The truth is not quite that simple. Figure 5.1 shows 95% CIs for the means of pairs of treatments. Details of the study don't matter here, only the degree of overlap in each pair of intervals and what it conveys about whether the two means are significantly different.

Take a minute and, using the overlapping intervals eyeball test, list for each pair whether the means are significantly different at the usual $\alpha = 0.05$. Go ahead. We will wait.

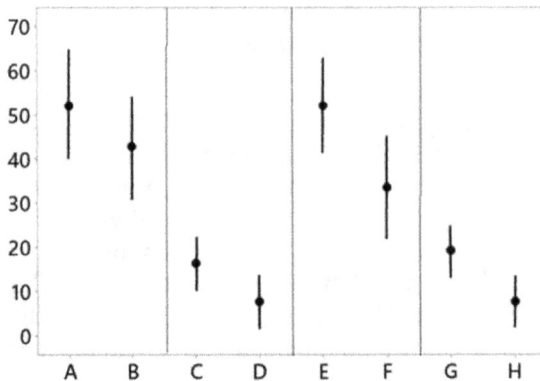

FIGURE 5.1
Four sets of pairs of confidence intervals, with varying degrees of overlap.

But not for long. Welcome back. We suppose you were clear about the first (not significant) and last (significant), but were perhaps less sure about the middle two tests. Here are the p-values, in order from left to right: 0.285, 0.045, 0.028, 0.007. As we see, the intervals can overlap somewhat, at least, and there can still be a significant difference between the two means.

Why is this so? For the sake of illustration, let's suppose that for both samples, the SE of the mean equals 1, in whatever units we are working with. Then for a 95% interval, the margin of error is approximately $2 \times SE = 2$. If the two means are further apart than 4 SEs, the two intervals will not overlap, and a p-value for the corresponding test will be small. Otherwise, the intervals will overlap.

BUT... the SE of the difference is: $SE(D) = \sqrt{SE_1^2 + SE_2^2} = \sqrt{1+1} = \sqrt{2} = 1.414$. And a 95% CI for the difference will have a margin of error of approximately $2 \times 1.4 = 2.8$. If the two means are more than 2.8 standard errors apart, the CI for the difference itself will not include zero. That is less stringent than insisting that they need to be 4 units apart. Intervals for two means can overlap somewhat and the difference between still be significant. Caution is required in using intervals in this way. *If you have access to the data, and if you wish to use intervals, it would be better to calculate the interval for the estimated difference and note whether it includes zero.*

5.4 One-Tailed Tests and Intervals

These days, a one-tailed test is often accompanied by a one-sided CI in many statistics packages. Recall that a classical two-sided interval captures the middle 95% of the distribution of the relevant statistic and yields a lower bound (LB) and an upper bound (UB). A t-based interval isolates 2.5% in each tail. In the mouse lifespan study, recall that the one-tailed test looking for an increase in lifespan above 31.3 yielded a p-value of 0.0225, a modestly significant finding against an alpha of 0.05. Here, a 95% LB is 31.6; it isolates the lower 5% of the distribution. We would say that we are 95% sure the true mean lifespan is greater than 31.6. This idea of one-sided intervals has not been in play in science for all that long, and most folks expect a confidence statement to take the form (LB, UB). Cultural norms take a long time to change. If you do a one-tailed test and accompany your report with a classical (i.e. two-sided) CI, it is likely that no harm will come from it, despite some risk of a mismatch between the test result and the inference from the CI. If your p-value is very tiny or very large, any mismatch will be virtually undetectable. With a marginal p-value, though, there is a risk of the estimate and test seeming to point to different conclusions.

5.5 Soap Box Comments on Hypothesis
Tests and Confidence Intervals

There are four things that make teaching students about hypothesis testing challenging.

1. The p-value approach is mathematically lovely and has greater consistency than the old critical t method[3] and greater utility[4] also, but it is based on what many practitioners view as arcane mathematics. We fear that too many students go on to use it without fully understanding it. And that's not their fault.

2. The conventional way of reporting hypotheses is backwards from the way they logically must be created. This adds an unnecessary point of difficulty for beginners. We expounded on that in Chapter 4.

3. A hypothesis test is by nature binary (reject the null or fail to do so), much like a criminal court case (found guilty or not found guilty[5]). That can sometimes make sense for tests done in a management setting if a decision has to be made based on the test, but science often does not depend on binary outcomes to move forward.

4. Done with care, proper estimates can often replace the information in a hypothesis test and add value (i.e. an actual estimate of the effect size or slope or whatever). Let's concentrate on this last point.

Suppose you make a 95% CI for the difference between two means or, say, the slope in a regression model.

Case 1. CI: (0.25, 6.25). The CI barely misses including zero, and so you could conclude that the p-value for the associated test would be just a little smaller than 0.05.

Case 2. CI: (0.25, 0.29). Here, relative to the scale of the interval, 0 is far away, and you could correctly deduce that the associated p-value would be tiny.

Flipping that around, a p-value of 0.0005 doesn't give you any clue about the magnitude of the effect. Even a scientifically irrelevant effect can be declared as statistically significant with a very small p-value, provided the sample size is big enough. For example, the center of CI one is 3; in CI two, it is 0.27. Of course, a lot depends on measurement units and context, but all else being equal, in Case 2, the effect is quite small compared to that in Case 1. Statistically significant is not the same as significant... Small p-values can arise from large effects, but also from small effects in studies with large sample sizes.

This sentiment about hypothesis tests is not ours alone. Ken surveyed a couple of dozen colleagues to get their take on what you just read. There was general agreement that including effect sizes and statements of precision[6] in journal articles has risen over the past couple of decades in prominence and has, in some cases, superseded reporting results of hypothesis tests.

Each discipline operates in its own "cultural space," and for some of them, *p*-values are passé, while for others, they are standard fare.

It appears that we are mid-stream on a paradigm shift on the subject. Paradigm shifts don't happen overnight and are often met initially by resistance. The survey respondents reflected that. Some of them have dispensed with test results in favor of properly reported effect sizes. Some agree with the premise but still like to see *p*-values. They all agree, however, that the era of simply reporting test results has gone by the wayside.

But don't take our word for it. Rod Little lays out a very elegant discussion of the *p*-value debate (Little, 2025, section 3.3). In that section, he references recent statements from the American Statistical Association (Wasserstein and Lazar, 2016; Wasserstein et al, 2019) that are in line with what we have written here. Indeed, those articles informed and sharpened our views.

5.6 Chapter Summary

We discussed the importance of choosing complementary alpha levels for tests and confidence levels for intervals. We shared some cautionary notes and suggestions for interpreting overlapping CIs. We pontificated on the place in statistical analyses for tests versus estimates (i.e. CIs).

Notes

1 Pretty darn sure...
2 This part is correct...
3 If you don't know what this is, don't worry. It will not come back to haunt you.
4 The utility is that a read can impose their own alpha-level. It is, after all a choice. The right alpha in one situation might not be suitable in another. Simply compare the *p*-level to *your* choice of alpha.
5 Sometimes people say, "found not guilty" but that implies "found innocent", which is not quite correct.
6 Either as a confidence interval or more simply by reporting the standard error of the estimate.

References

Little, R. 2025. Seminal Ideas and Controversies in Statistics. Chapman and Hall, 223 pages.

Manly, B.F.J. and Navarro Alberto, J. 2021. Randomization, Bootstrap and Monte Carlo Methods in Biology. (4th Edition). Boca Raton, FL: CRC Press.

Scheaffer, R.L., W. Mendenhall III, R. L. Ott, and K.G. Gerow. 2012. Elementary Survey Sampling, (7th Edition). Cengage Learning, Boston MA. 436 pages.

Wasserstein, R.L. and N.A. Lazar. 2016. The ASA's statement on p-values: context, process, and purpose. American Statistician 70(2): 129–133.

Wasserstein, R.L., A.L. Schirm, and N.A. Lazar. 2019. Moving to a world beyond $p < 0.05$. American Statistician 73: 1–19.

Part II

Introducing Regression Models

Many analyses that researchers do will involve trying to connect the dots between a collection of explanatory variables and some response variable. Here, we focus on a numerical response variable. The collection of predictors might include numerical and categorical variables.

We will begin with simple linear regression, in part because a key element to doing such analyses is checking the assumptions to ensure that the resulting model is valid. Luckily, this process is virtually identical for more complicated models, so studying that here, in the context of a model with a single numerical predictor, is quite convenient.

Chapters 7 and 8 study extended examples and share deeper considerations of regression models. Chapter 9 studies the use of logarithmic transformations in some detail. We take the view that this tool (log transformations) is not mandated by skewed data but is rather very useful in capturing relativity in relationships. For example, in a case of exponential increase, the average value of the response increases by some percentage for each specified increment of increase in the predictor.

DOI: 10.1201/9781003609605-7

6

Introduction to Simple Linear Regression

Simple linear regression[1] (SLR) is often used as the starting place for the study of more complicated models. In research, it is common to have several predictors or factors under consideration; some of them may be numerical, others categorical. However, the process of implementing SLR (checking validity conditions, interpreting output, etc.) will carry over very directly into that more complicated setting. So, it is a reasonable place to start learning about implementing and using statistical models.

What will you learn here? This chapter is meant to provide intellectual comfort to you. Valid use of the fun stuff (testing hypotheses, making estimates and predictions) from regression models depends on adequately meeting the necessary assumptions. This chapter is devoted to introducing those assumptions to you and showing you, by examples, how to assess them. Assessing them requires judgment, which in turn requires practice, which we will give you here.

What do you need to know already? There are several topics that will come here that are covered in earlier chapters of this book. These include:

- Standard deviation and standard error,
- Central Limit Theorem, which addresses the role of the Normal distribution,
- Hypothesis testing concepts and procedures, and
- Confidence interval concepts and properties.

What's novel? It might surprise you to learn that there are no distributional assumptions about the predictor and the response. Neither needs to have a Normal distribution. In fact, the assumptions mention neither of them, aside from needing them to be meaningfully numerical.

6.1 A Motivating Example

We are interested here in the timing of eruptions of the famous "Old Faithful" geyser. National Park Service interpretive naturalists stationed in the Upper Geyser Basin are often approached by visitors wanting to know how long

DOI: 10.1201/9781003609605-8

they must wait until the next eruption. Predicting that time is our goal. We are using this dataset for several reasons:

- The response and predictors are easy to understand; they do not depend on knowledge of any specific area of study.
- It has some distributional features that are instructive when considering the assumptions behind regression models.
- In almost any statistical modeling situation, you will have to make choices. Do you use a simpler model, with perhaps lower ability to predict, but increased ease of explanation? Is a more sophisticated model desirable with better predictability at perhaps the price of elegance?

The dataset we will use has many sources for variations on the actual data[2]. What is likely obvious to you: the data do not have a Normal distribution (Figure 6.1). Further, you can see it is difficult to make a sharp prediction. If you want to cover the range of "most likely" waiting times, you might have to say between 42 and 95 minutes. That is not a narrow range; historically, times range from about 30 to 120 minutes. The mean is approximately 71 minutes, but declaring that as a prediction would be ill-advised, given the amount of variation. This situation is unsatisfying to both the naturalists and their visitors.

Fortunately, there is hope. As it happens, the duration of the most recent eruption is a good predictor of the current waiting time: shorter durations being predictive of shorter waiting times, and longer of longer waits. Park naturalists use this relationship to great advantage. We will now formally study that relationship as an introduction to SLR. Let's begin with a scatterplot (Figure 6.2) of the data and an introduction to some terms and conventions.

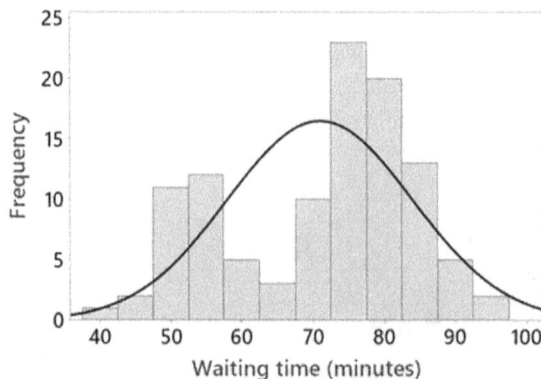

FIGURE 6.1
Histogram of waiting times between eruptions of Old Faithful Geyser.

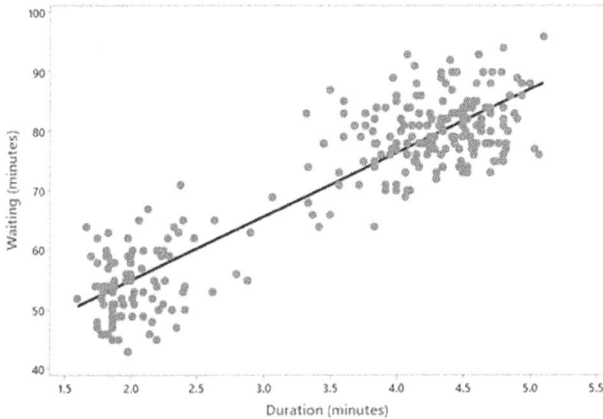

FIGURE 6.2
Scatterplot with fitted regression line for waiting time versus duration of previous eruption.
Line equation is $W = 33.47 + 10.73 \times D$.

In the formulation of the regression line, 33.47 is called the intercept, and
10.73 is the slope. We will discuss these in more detail shortly. As an abstraction, we would write this as $b_0 + b_1 X$; the reason for starting the subscript
counting at 0 is not important here, but that *is* now our cultural convention.
You might have seen the formula for a straight line in high school math or an
algebra class, as $mX + b$. Same idea, different symbols, and a different order
of presentation.

In this case, the duration of previous eruption (D) is the predictor or explanatory variable, while waiting time (W) is the response variable. Generally,
"Y" denotes the response variable, and "X" denotes the predictor. Some of
you might know of X as the independent variable, and Y as the dependent
variable. Those terms are not wrong but have fallen somewhat out of fashion. The terms "response" (or "outcome") and "predictor" (or "explanatory")
more clearly indicate the role taken by each, since we are interested here in
predicting waiting time, these choices are natural. If, for some reason, we
were interested in predicting the length of an eruption, we might reverse the
roles of D and W. Which variable is the response, and which is the predictor
are story-telling choices.

It is usual to draw the scatterplot with the predictor on the horizontal axis,
and the response on the vertical one. This is merely a convention, but if you
stick with it (and so does everyone else), reading such graphs can become
relatively easy. For example, suppose the most recent eruption lasted four
minutes. Read up from $D = 4$, and you will see the waiting for the next eruption will likely be between approximately 70 and 90 minutes. This is a quite
narrower and shifted range than from 42 to 95 minutes!

The regression line is an attempt to estimate the mean of Y for given values
of X via a straightforward mechanism, namely a straight-line model. We will

explain the thinking behind the equation for the line $(33.47 + 10.73 \times D)$ in the next chapter, but for now, note that in Figure 6.2, for any value throughout the observed range of eruption durations, the line divides the waiting times into approximately equal parts. That is, for a given duration, about half the waiting time values are above the line, half are below. Before we learn how to use this model, it behooves us to consider the conditions under which the straight-line model shown here is appropriate.

6.2 Validity Conditions for Simple Linear Regression

We will list the assumptions without comment here, and then study each in turn in more depth, and apply that learning to the geyser data. The order in which we list them here is important, as we believe that considering them in this order is helpful. For instance, if any of them fail, there is no point in going on to consider the next ones. Stop. Fix the problem and *then* move on. The details of the tools you will use to assess these depend on which statistics package you use; we will keep this discussion free from being tied to one or another.

If the assumptions are reasonably well met, then you can trust the resulting output. Assumption checking can also guide you towards model improvements, more on which when we get to multiple regression. Once you feel comfortable assessing the assumptions for SLR, you are done. The assumptions required for more complicated models are essentially identical to those for SLR; even better, assessing them is identical. What a deal!

List of assumptions:

0. Both Y and X need to be meaningfully numeric; the data come in pairs: each experimental or sampling unit yields up a Y and X; each pair needs to be independent of other pairs. You cannot necessarily test assumption (0); it is something you need to know about the data before you begin a regression analysis.
1. The model (a straight line here) is a good fit.
2. The amount of variation around the model is constant throughout the range of the predictor[3].
3. The distribution of the residuals is Normal enough.

6.3 Checking the Model Assumptions

Checking the assumptions is done by examining certain features of the model residuals. So, let's start by introducing them. "Residuals" is an important term to understand when discussing regression models, and so we will

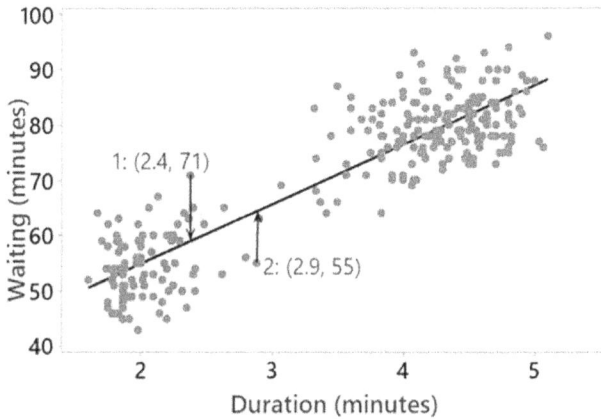

FIGURE 6.3
Reprise of Figure 6.2, with two data points highlighted to illustrate residuals. The arrows point to the fitted values for each.

illustrate them using the geyser data. Each point in Figure 6.3 represents a pair of values (W, D), where W is the waiting time before an eruption and D is the duration of the previous eruption. There is a third value we can associate with each point, namely the fitted value of W, given D. To differentiate this from the observed response, it is denoted $\hat{W}$.

Some of the observed values of W are quite close to the fitted line, others not so close. The differences ($W - \hat{W}$) are called the residuals. There is one such value for each point on the graph. For instance, given $D = 2.4$ minutes, and $W = 71$ minutes (point 1), the fitted value is $\hat{W} = 33.5 + 10.7 \times 2.4 = 59.3$, yielding a residual of $71 - 59.3 = 11.7$. As a second example, for $D = 2.9$, $W = 55$ (point 2), and $\hat{W} = 33.5 + 10.7 \times 2.9 = 64.6$, yielding a residual of $55 - 64.6 = -9.6$. The average of all the residuals is precisely 0, by construction of the line.

We imagine that you are keen to get to the answers: how to do tests and make estimates using regression. However, you should check the assumptions before looking at the numerical output; you need to know whether said output is valid. So let's step through them one at a time and use the geyser data to exemplify the analysis.

6.3.1 Assumption "0": Data Requirements

Let's begin with (0), and, of course, the first question on your mind is why "0"? Why not "1"? If you have simply been given a data set to work with, often the case in statistics courses, it's not easy to test this one. You must know where the data came from. If it is your data, from your research, you will of course know whether the data meet the assumption of meaningful numeric. That being said,

1. Watch out for numbers being used as labels. For instance, if X was color, and you labelled the five colors in the study as Color 1, 2, and

so on, then this would not be a suitable data set for regression, since the predictor is not meaningfully numerical. The numbers are simply labels for categories.

2. An example of failure of independence: suppose you had data from 24 piglets, with age as the predictor and weight as the response. If, say, they came in sets of four from each of six litters, they would not be mutually independent data points since you might expect two piglets within a litter to be more like each other than any two from different litters.

3. Another common example of failure of independence is when you have observations on each subject repeated over time. We tackle this issue explicitly in Chapter Seventeen.

Notice that in this discussion of the origins of the data, we did not mention the need to have a random sample. That was intentional. Imagine the following experiment. You are studying growth rates of plants in growth chambers. You have six growth chambers with the temperature set at 35°F. And other sets of six at each of 45°F, 55°F, 65°F, 75°F, and 85°F. The predictors in that case have no random distribution whatsoever; you had chosen them. This would still be a viable data set for which a regression model would be worth considering. So, the distribution of the predictor is moot, at least as regards basic assumptions. What about Y (whatever has been chosen; it might be biomass after some weeks)? There will be a clump of values centered around whatever the average response is for 35°F, another for the response at 45°F, and so on. It is fair to say that the response, given a value of temperature, should be random. That is an easy assumption to satisfy.

Discussion of regression often includes the phrase, "...conditional on X..." That phrase is jargon for what you just read, namely that there are no assumptions about the distribution of the predictor, which implies there are no assumptions about the distribution of the response. It is only the residuals that matter. The geyser data are a good real-life example of this.

6.3.2 Assumption One: Goodness of Fit

This one is counterintuitive; easy to assess, but it takes some "soak time" for true understanding. Goodness of fit (GoF) for a regression model is not an assessment of how well the model works. A significant regression (i.e. the slope is clearly non-zero) and/or a high R^2 (more on which later) are no guarantee of a good fit in the sense meant here. A GoF test answers the question: with the given predictor and model, do we need to use something fancier, which might include a quadratic term, or perhaps employ transformations, or is the posited model adequate in that there are not suitably beneficial improvements?

Real data sets are often messy, so here are four artificial data sets, each chosen to more sharply make the relevant points. Figures 6.4 (panels a–d) and the subsequent discussion lay out the situation.

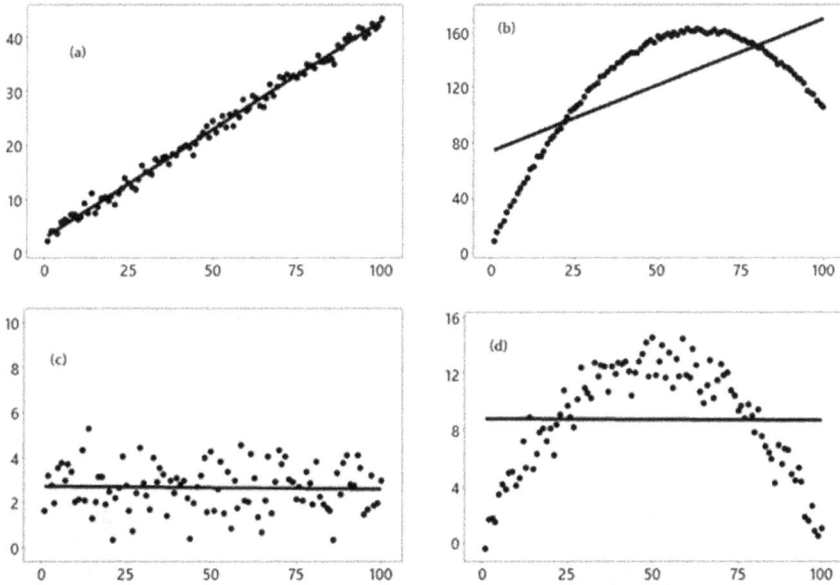

FIGURE 6.4
Illustrations of goodness of fit (not all are good).

Discussion:

- It seems quite clear that in panels (a) and (c), a straight line fits the data adequately, meaning no improvements are obvious.
- Panels (b) and (d) show distinct curvature (i.e. a straight-line model is demonstrably not a good fit). These would not pass a GoF test for a simple linear model.
- The numerical output from a regression for panels (a) and (b) would declare a significant relationship: a straight line fit clearly has a non-zero slope.
- On the other hand, the data in panels (c) and (d) would show no simple linear relationship. Dismissing the idea of a relationship would be a mistake for (d); simply, that relationship requires curvature.

To sum up, a straight-line model might be an adequate fit (a and c) or not (b and d). Quite distinct from that question, a straight-line model might appear on the surface (i.e. based only on the numerical output, to do a good job (a and b) or not (c and d).

6.3.2.1 Goodness of Fit Test for the Geyser Data

GoF is easily assessed by examination of the Residuals versus Fits plot. Before we get to that, let's examine where residuals come from, for the case of the

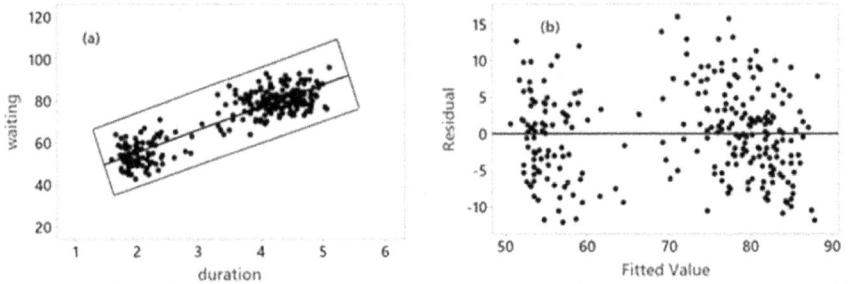

FIGURE 6.5
Scatterplot of geyser data, with the data "boxed in" (a). Resulting Residuals versus Fits (b).

geyser data. In Figure 6.5a, we added a box around the data and fitted line to demonstrate the origins of the Residuals versus Fits plot (Figure 6.5b). It is essentially a graph made from the values in that box, rotated so that the fitted line is now a horizontal line, with each observed value located directly above or below its fitted value. A datum on the fitted line would have a residual value (observed minus fitted) of zero. For many real-life situations, the Residuals versus Fits plot highlights relevant potential curvature or unequal scatter that cannot easily be seen with a scatterplot of the original data. More importantly, for models with more than one predictor, this plot is invaluable because it takes a high-dimensional set of data and reduces the relevant information to a simple and easy-to-use scatterplot.

By construction, the average of the residuals is zero. Scan Figure 6.5b from left to right; is the average of the residuals approximately zero throughout, or does there appear to be distinct curvature? That is, does the middle of the cloud of residuals systematically go up and then back down, or the reverse (down, then back up)? There is a suggestion of curvature in this case, but it is pretty subtle. There is no concerning lack of fit, and so we declare that the GoF is "good enough."

6.3.3 Assumption Two: Homoscedasticity

If the amount of scatter around the regression line is not constant, the formula for the SE of the slope and that for the intercept are invalid, since they are built on an estimate of that supposed constant variance. Also, we cannot speak of the residuals having *any* single distribution. We cannot ask if they are approximately Normal: if the size of the variance keeps shifting, then there is no one distribution...

We use the same graph for this as we used for the goodness of fit test (Figure 6.5b). The amount of variation, depicted here as the vertical scatter, is relatively consistent as we look from left to right through the range of fitted values. Done: equal scatter is OK. When this assumption fails, it usually does so in a dramatic and often predictable fashion, which we will illustrate once we are done introducing the assumptions.

6.3.4 Assumption Three: The Distribution of the Residuals is Normal Enough

This is a Central Limit Theorem moment. If the model is a good enough fit, and if the variation about the line is approximately constant, then if the residuals are Normal enough, the distribution of the slope and intercept coefficients is approximately Normal, and the usual tools will work for testing and estimation. The word "enough" calibrates for sample size. In the case of the geyser data, the sample size is so large that we are almost guaranteed that the distribution of the two model coefficients is close to Normal, which makes the tests and confidence intervals valid. Figure 6.6 shows a histogram of the residuals for those data.

These data do not have a Normal distribution; they are somewhat skewed right[4]. The Central Limit Theorem says that the regression coefficients will have at least an approximate Normal distribution if the sample size is large enough, without much regard to the underlying random distribution. Due to the very large sample size ($n = 272$), we are confident that the distribution of the intercept and slope estimates will indeed be approximately Normal. The distribution of the residuals is Normal *enough*.

6.3.5 Assumptions Regarding Y and X

There. That does it. WAIT (you are thinking)! We said nothing about Y or X having a Normal distribution. In fact, we did not mention them at all! We mentioned that issue in our listing of the assumptions, but the notion that the data need to be Normally distributed is so pervasive that my comments

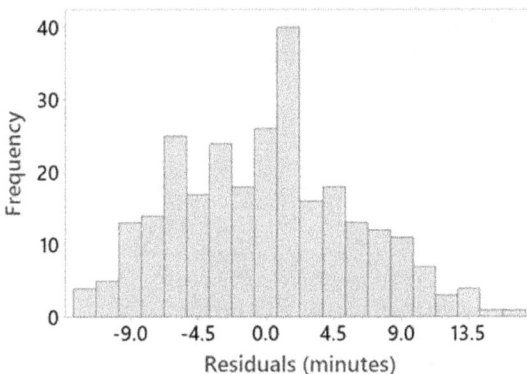

FIGURE 6.6
Distribution of the residuals for the geyser data.

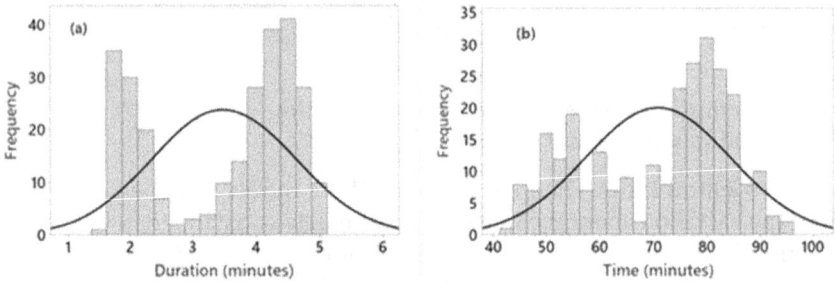

FIGURE 6.7
Histograms of predictor and response variables for the geyser data. (a) Duration of eruptions;
(b) waiting times.

there might not have stuck. The geyser data are an interesting case in point
for that discussion. Let's start with a histogram of each variable (Figure 6.7).
  Whoa! Hugely not Normal. Let's think about this one step at a time.

  1. The eruption durations distribution is clearly bimodal (panel a), with
     a smaller collection near 2 minutes of duration, and a larger, more
     spread-out bump centered just above four minutes.
  2. Now let's imagine that there is a reasonably straight-line relationship
     between the average waiting times and duration, and that there is
     equal scatter among the residuals, and that the distribution of the
     residuals is adequately Normal. We have that in this case.
  3. Given all that, we would expect the waiting times to have a cluster
     of smaller times and another larger cluster of longer times. So it is
     (panel b).

  You can see all this play out in the scatterplot of the data (Figure 6.2). You
can clearly see the bimodal distribution of geyser durations. And you can
also see the collection of lower waiting times as well as the larger collection
of longer times.
  The utter failure of *Y* and *X* to be Normally distributed is utterly irrelevant.

6.4 Testing the Assumptions: Examples

We both recall the start of a second semester of applied statistics, which would
begin with, "Let's warm up with a straightforward multiple regression. Class,
what are the assumptions we need to test?" They would mostly stare at us,
paralyzed. Simply delivering the list of assumptions and testing them once

or twice did not suffice to deeply understand them. It can be difficult to judge when they are acceptably met, and what to do when they are not.

In fact, assessing the assumptions and using them to guide model construction requires judgment. This in turn requires much practice[5]. In our experience, beginners are often too rigid in assessing them. It takes time and practice to get comfortable in asserting your judgment, in declaring, "good enough." We want to go through a variety of examples here to help you along this path. We will focus here on GoF and equal scatter, and, if those assumptions pass muster, then a histogram of the residuals to see if their distribution is Normal enough. We will refer to the process of checking these two plots as "examining the residual plots." For many of the examples, we will not discuss the model itself, to keep our focus on the assumption testing.

Sometimes everything is perfect.

Frances Galton (1886) studied the heights of fathers and their sons, and observed a phenomenon he called "regression to the mean," which led to this analysis method being called regression. We will discuss that phrase in Chapter 8 to show you a way to make sense of it. Here, though (Figure 6.8), we will show you the residual plots for the regression of the heights of sons versus those of their fathers.

This is a relatively rare example. Enjoy it. The relationship is quite strong, and a straight line is a good fit with equal scatter (panel a). The distribution of the residuals is quite Normal-looking (panel b). Two comments, though.

1. The residuals versus fits plot displays the ideal "random cloud of dots." There are fewer data points on the left (shorter fathers) and right (taller), so the data points at each end are not quite "filled out." This is usual.

2. This clean-looking pattern is due to the large sample size; it is over 1000. Random things behave more predictably when there are more of them. As you will see below, you might not expect such clean patterns with smaller data sets. Be gentle with them.

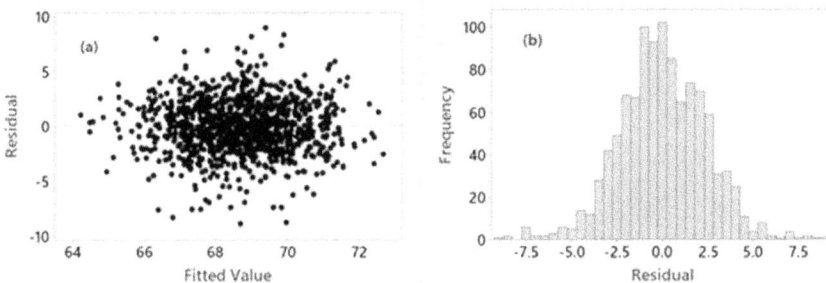

FIGURE 6.8
Residuals versus Fits plot (a) and histogram of the residuals (b) for the father/son heights.

6.4.1 Sometimes Oddities are Quite Harmless

In this example (Johnson and Courtney, 1931), cited by Singer and Willett (1991), young children were asked to stack wooden blocks; some were cylinders, others cubes. Of interest was how many they could stack, as a function of the age of the children and the shape of the blocks. We won't pursue the model here, but take note of the Residual versus Fits plot (Figure 6.9).

GoF is OK, but equal scatter is not. That being said, what is eye-catching is that series of diagonal lines of dots in the residuals versus fits plot. That is weird. Interestingly, it is merely an artifact of the response variable being small integer values. The response values range between 3 and 13, with the vast majority being between 4 and 9.

How about this for oddities?

To investigate the effect of caffeine on performance on a simple physical task, thirty male college students were trained in finger tapping (Draper and Smith, 1981). They were then divided at random into three groups of ten, and the groups received different doses of caffeine (0, 100, and 200 mg). Two hours after treatment, the number of taps per minute was recorded for each subject. The residual plots are in Figure 6.10.

It might seem odd that all the residuals are neatly stacked into three columns and, within columns, seem mostly evenly spaced. This was a designed experiment with three fixed values of caffeine, which leads to only three specific fitted values. Hence the columns. And the fact that the response values are integer valued (number of taps) explains the mostly equal spacing. No harm. There is no evidence of curvature and no evidence of unequal scatter. The histogram of the residuals is reasonably symmetric.

Be careful about the orientation of "equal spread"

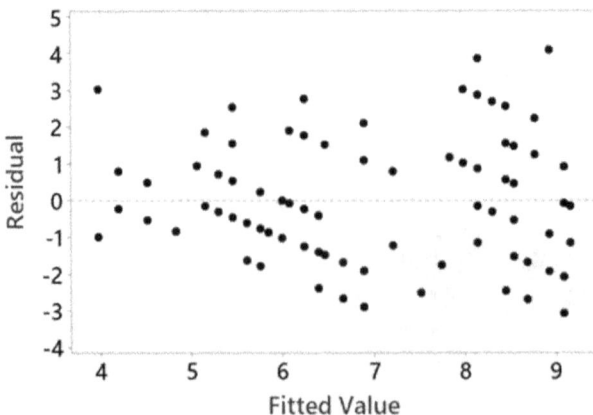

FIGURE 6.9
Residual versus fits plot from a regression modeling the number of blocks children could stack as a function of their age.

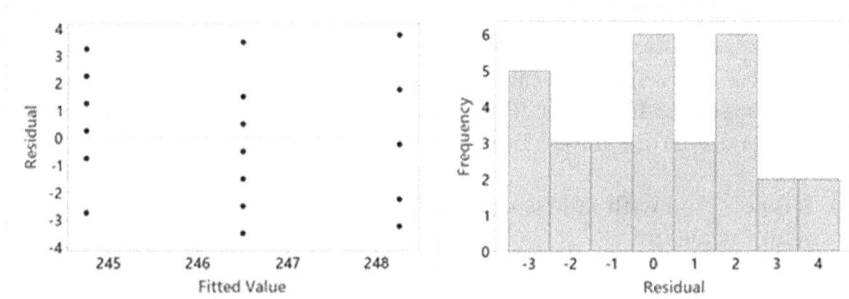

FIGURE 6.10
Residual plots for the caffeine study.

These 32 data values were collected from cross-country races, with the response being the record winning time and the predictor being the length of the race (Hand *et al*, 1994, page 273). As you might expect, times go up as race length goes up. Residual plots are in Figure 6.11.

When faced with these plots, students immediately noticed that the values are not equally scattered from left to right: they are clustered to the left. True enough, but also not a problem for the regression assumptions. It merely reflects that the distribution of the predictors is skewed: a large number of shorter distance races and a few longer. That is all. No harm done. The distribution of the predictor is not a feature in the regression assumptions. That said, you might be cautious about making predictions for longer races, since there is not much data for them.

Also, some might see curvature in the residuals versus fits plot. If you can change your judgment about curvature, or unequal scatter for that matter, by covering up (or mentally adding) a single data point to the plot, what you are seeing is likely not a problem. For example, one more datum at about Fitted = 110, residual = 20 would make the apparent curvature vanish.

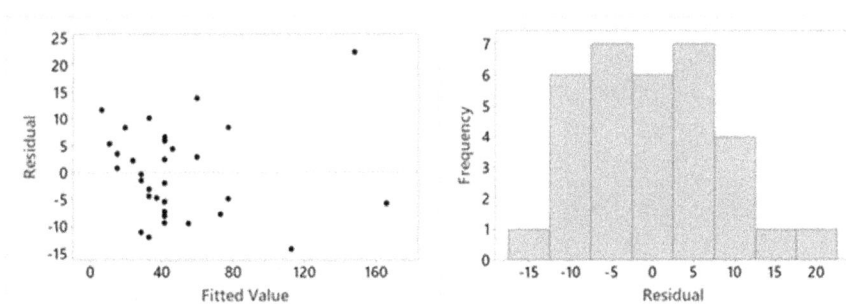

FIGURE 6.11
Residual plots from a regression of time of race winner against length of race.

The next example focuses on unequal scatter, in addition to a lack of GoF. When unequal scatter occurs, it is usually not haphazard. Typical failures of equal scatter show smaller scatter for smaller fitted values and systematic increases in scatter for larger fitted values. This is due to two often-related phenomena occurring at the same time:

1. It is common with ratio scale data that variation increases in tandem with the means.

2. Relativity in relationships is also common. For instance, the response might change in a relative way (some percentage increase or decrease) for every unit change in the predictor. Or a relative change in a pre-dictor (e.g. doubling or a 50% increase) might be associated with a constant increment of change in the average of the response. Or, not uncommonly, relativity can exist on both axes.

We will study this matter more deeply in Chapter 9 (on log transforma-tions), but here we will show you how examining the residual plots can give you relevant clues.

Hormone levels over time

Huber *et al* (2024) studied levels of regulatory and functional adrenal gland proteins across the life course, using baboons as a model. Citrate synthase (CS), measured by visible presence (% area) in stained adrenal gland cells, a measure of mitochondrial function, was one of their response variables. The data we use here are adapted from that work. In Figure 6.12, we see a SLR fitted to those data. The model is discomforting for two reasons. One is that it appears that for animals over twenty-two, the predicted values would be negative, which, of course cannot happen. Second, the scatter around the line is much greater for younger animals who have higher CS levels, on average, than do older animals.

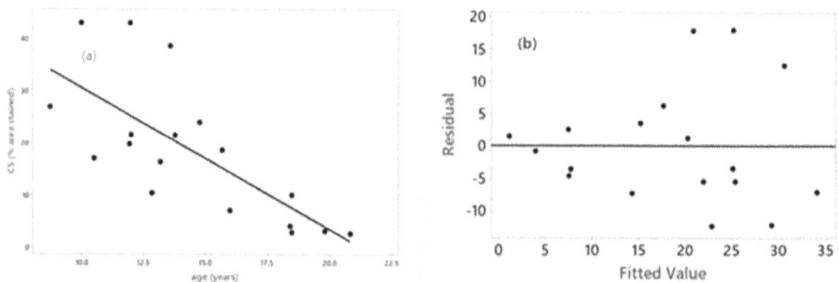

FIGURE 6.12
CS levels versus age: fitted regression (a; $CS = 57.75 - 2.72 \times A$) and residuals versus fits (b).

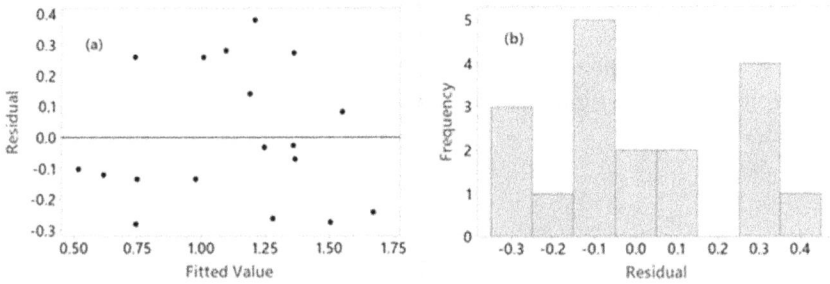

FIGURE 6.13
Residual plots for log(CS) regressed against age.

Simultaneously, this model shows a distinct lack of fit and unequal scatter: the average of the residuals drops into the middle of the plot (reading from the left of panel b) and appears to be headed back up. Also, the variation is clearly larger in association with larger fitted values. This pattern (curvature down, then back up, accompanied by more scatter in association with higher predicted values) is characteristic of data where the response changes in a *relative* way rather than an incremental way. Given the failure of GoF and equal scatter, there is no point in looking at the histogram of the residuals. The current situation shows quite clearly that there is no single distribution for the residuals, in which case there is no "it" to be Normal.

The fix to this is to use a logarithmic transformation on the response variable. Skipping details for now, Figure 6.13 shows the relevant residual plots.

Patterns in random things get clearer with larger sample sizes. With small sample sizes, you should allow some grace for flakiness. Given that, we declare these residual plots adequate.

For the sake of tying up the picture a bit more, Figure 6.14 shows the fitted line when CS values have been log-transformed. Panel (a) shows the data on

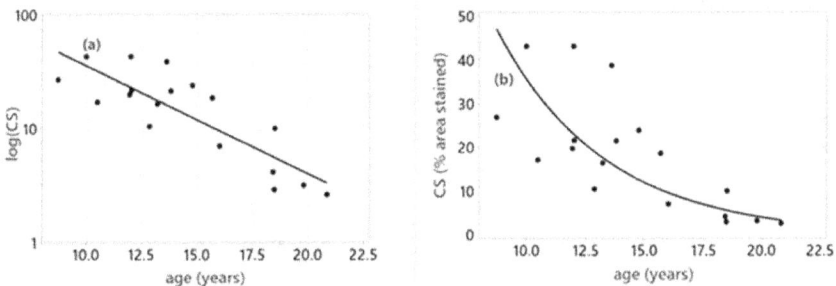

FIGURE 6.14
(a) Fitted line $\log(CS) = 2.4 - 0.095 \times A$, on the log-scale, (b) back transformed to the original scale.

the log scale. A straight-line model seems plausible, and that the scatter is relatively constant). Panel (b) shows the model back transformed to the original scale. We will come back to these data in Chapter Nine.

If you wish to investigate more examples of how to use residual analyses, see Dodge (2008).

6.5 Chapter Summary

We introduced the basics of regression by way of introducing SLR. We listed the necessary assumptions that must be met, at least approximately, for a valid regression model. In our view, it is useful to check these assumptions in a certain order because failure at any one step means you should stop and reconsider the model. Assuming your response and predictor variables are meaningfully numerical and suitably independent, this order is to check for GoF, then equality of scatter in the residuals, and finally, whether the distribution of the residuals is Normal enough.

Both GoF and equality of scatter can be easily assessed by examining a plot of residuals versus fits. GoF is a simple idea, but an intuitive reading of the phrase can mislead the researcher into failing to check it properly. The residuals do not need to have a Normal distribution; the word "enough" is a trigger word to get you to make the judgment with the sample size in mind. The larger the sample size, the less perfectly Normal the residuals need to be. This is because the Central Limit Theorem will bring approximate Normality to the distributions of the regression coefficients, thus making inference via t-tests and t-based confidence intervals valid.

Perhaps a surprise in the chapter is that it is NOT an assumption that your response variable has a Normal distribution. Ditto for the predictor.

Learning to use the assumptions won't happen simply by being able to list them and seeing one or two examples. For that reason, we ended the chapter by studying examples wherein careful examination of the assumptions can lead to improvements in the modeling.

6.6 Exercises

1. In a study to determine the impact of argument on jury decision-making, psychology students were asked, after watching a video of a trial, to render a verdict (called **Pre-verdict**, Figure 6.15) as follows:

   a. **0** for "no verdict",

   b. **1** for 1st degree murder,

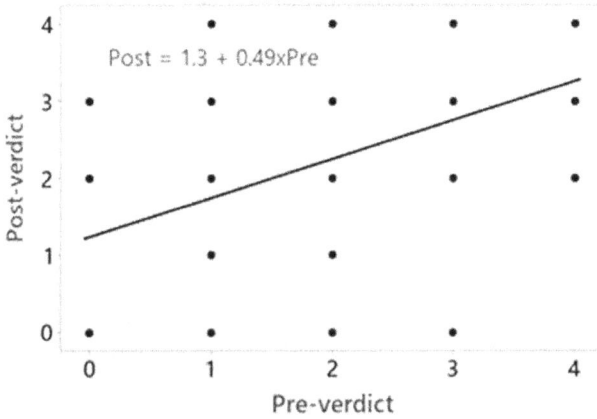

FIGURE 6.15
Verdicts before versus after a debate on a criminal case.

    c. **2** for 2nd degree murder,

    d. **3** for manslaughter, and

    e. **4** for not guilty by reason of self-defense.

After a debate of the merits of the case, they were asked to render a second verdict (**Post-verdict below**) to see if the debate affected their decisions. Are the regression assumptions met? Aside: there were many students in the experiment, and so each dot in Figure 6.15 might well represent many students.

2. In a study to establish a relationship between weight and length of black bears, the following regression (Figure 6.16) was fitted to the data.

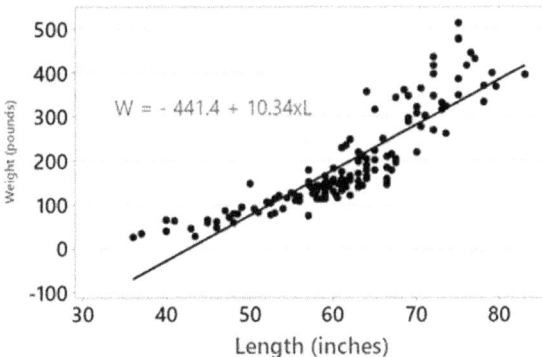

FIGURE 6.16
Regression of weight on length for black bears.

Are the regression assumptions met with these data?

3. Ignoring your answer to the previous question, according to this model, by how much does weight increase on average for each inch of extra length?

4. Ignoring your response to question 2, what problems are caused by the intercept (−441.5) being so hugely negative?

6.7 Solutions

1. These are not actual numbers; they are labels for ordered categories. As such, regression is not a suitable tool for analysis.

2. The data are meaningfully numerical, but curvature in the relationship and increasing scatter for higher weights are readily visible. So, no.

3. This question asks you to interpret the slope. For each one unit (an inch) increase in the predictor, weight increases by 10.34 pounds, on average.

4. This causes no problems. Formally, the intercept is the estimate of the average weight for a length of zero, in which case there is no bear!. The range of interest for this model is length between 35 and 85 inches.

Notes

1 "Simple" here only implies that there is a single predictor, and so the adjective stands in contrast to "multiple" (coming soon to chapters near you).

2 The copy used here came from online resources associated with W. Härdle's 1991 textbook, "Smoothing Methods with Implementations in S..

3 The so-called "homoscedasticity" (i.e. equal scatter) assumption, if you want the $10.00 word for it.

4 It you are inclined to call this distribution skewed left, check our Statistics-English translation dictionary at the end of Chapter Two.

5 "Good judgment comes from experience, and a lot of that comes from bad judgment." This aphorism is variously attributed. Many sources indicate it was first uttered by American humorist and astute observer of life Will Rogers.

References

Dodge, Y. (2008). *Residual Analysis*. The Concise Encyclopedia of Statistics, pp. 5–9. Berlin: Springer.

Draper, N.R. and Smith, H. (1981) Applied Regression Analysis, (2nd Edition), New York: John Wiley & Sons, 425 pages.

Galton, F. 1886. Regression towards mediocrity in hereditary stature. The Journal of the Anthropological Institute of Great Britain and Ireland (15): 246–263

Hand, D.J., E. Daly, A.D. Lunn, K.J. McConway, and E. Ostrowski. 1994. A Handbook of Small Data Sets. Chapman and Hall, New York. 458 pages.

Hardle, W. 1991. Smoothing Methods with Implementations in S. Springer Series in Statistics. Springer-Verlag, New York. 262 pages.

Huber, H.F., C. Li, D. Xie, K. G. Gerow, T. C. Register, C. A. Shively, L. A. Cox, and P. W. Nathanielsz. 2024. Female baboon adrenal zona fasciculata and zona reticularis regulatory and functional proteins decrease across the life course. GeroScience, 46(3): 3405–3417. https://doi.org/10.1007/s11357-024-01080-9

Johnson, B. and D.M. Courtney. (1931). Tower building, Child Development, 2(2), 161–162

Singer, J.D. and J. B. Willett (1990). Improving the teaching of applied statistics: Putting the data back into data analysis. The American Statistician, 44(3), 223–230.

7

Regression by the Numbers: Making Sense of and Using the Output

Now that we are satisfied that the regression assumptions are reasonably well met for the geyser data, we can work our way through the numerical output, confident that it is valid.

With the output from fitting a regression model, you might want to

1. Do a hypothesis test regarding the slope. "Is there a significant relationship between the response and predictor?"
   Given that there is...

2. Explain how changes in the predictor are related to changes, on average, in the response. This might require a confidence interval (CI) for the slope.

3. Predict the average response or a yet-to-be-seen individual response.

Research in a scientific setting usually focuses on the first two. In a management setting, prediction might be of interest. The geyser data are an example.

7.1 The Regression Coefficients

You already know the fitted regression equation for the geyser data: $W = 33.47 + 10.73 \times D$. Conventionally, a statistics package will report the equation itself, a summary of the coefficients (Table 7.1), some overall summary statistics (Table 7.2), and an ANOVA[1] table (Table 7.3). We will follow each table with a discussion of its contents.

Let's pause to remind ourselves of what these intercept and slope represent. Historically, one would have drawn a scatterplot and the regression line on a piece of graph paper (Figure 7.1).[2] The first step is to draw the axes, setting them

TABLE 7.1

Coefficients from Fitting a Simple Linear Regression

Term	Coefficient	SE(Coeff)	t-Value	p-Value
Intercept	33.47	1.15	28.99	0.000
Slope	10.73	0.315	34.09	0.000

DOI: 10.1201/9781003609605-9

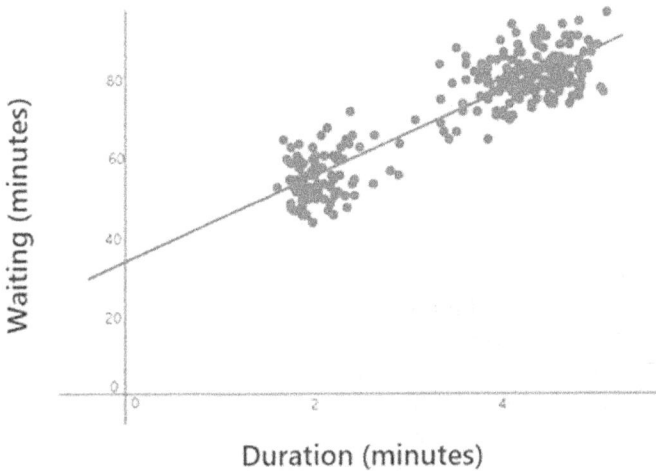

FIGURE 7.1
"Old school" scatterplot of the geyser data with fitted regression line.

to cross at the origin: $Y = 0$ and $X = 0$. A convenient way to draw a straight line on such a graph is to determine a point through which it runs and then draw it with the appropriate tilt (slope). In this context, the place where the line crosses the Y-intercept is a convenient place to "locate" the line. Then the slope can be brought into play (it is easy, for instance, to determine the height of the line at $D = 4$). We no longer draw graphs that way, but the choice of "intercept, slope" to characterize a regression line is now culturally ingrained.

Strictly speaking, the intercept estimates the average of the response when the predictor equals zero. Here, the range of interest for the predictor is from about 1.5 to 6 minutes. Relative to the scale of those numbers, zero is nowhere near the interval of interest. So, we usually do not care one way or another about the numerical value of the intercept.

The slope is meaningful, though. Here it tells us that, on average, the waiting time between eruptions is 10.73 minutes longer for every extra minute of duration of the previous eruption.

The slope is the basis for the classical explanation of the relationship between the response and the predictor. You will typically use a sentence along the lines of, "The mean of Y changes by b_1 for every one-unit increase in X." We will need to use both for making predictions (section 7.6).

7.2 Summary Statistics: R^2 and Its Relatives

It is usual for statistics packages to report the SD among the residuals as well as the coefficient of variation, R^2 (Table 7.2), which tells you the proportion

TABLE 7.2

Model Summary for the Geyser Data

s_{res}	R^2	R^2_{adj}
5.91	81.2%	81.1%

of the variation in the response that is attributable to the model. Here, $R^2 = 81.2\%$. We will come back to the adjusted R^2 below.

How big does R^2 need to be? Inevitably, when R^2 is introduced, students want to know how big it should be. What values are big enough to be considered good? What are too small? If we turn it back to them, their usual advice is to use 75% or so as a lower limit to "good enough." Let's suppose you are building a model for predicting the strength of concrete with differing proportions of various ingredients being the predictors. Neither Ken nor Jorge wants to drive on a bridge built with concrete where the developers could only explain 75% of the variation in strength. 95% or 99% might be more like it. On the other hand, if you reported a model trying to explain some aspect of human behavior and showed an R^2 of 75%, we might be suspicious that you did something incorrect, since explaining that much of human behavior would be astounding. In short, it depends on the context. Its adequacy will depend on your judgement.

On R^2 and correlation. The correlation between two variables measures the degree to which a scatterplot of the two would fall on a straight line. It takes on values between −1 and 1; it is negative when the slope of the relationship is negative, and positive when the slope is positive. For simple linear regression, R^2 is the square of the correlation, rescaled into a percentage. We find R^2 to be more compelling than correlation for discussion because of the definition: the proportion of the variation in the response that is attributable to the model. That statistic and its meaning will carry over directly into models with multiple predictors, whereas correlation does not.

This statistic takes on values between 0 (no correlation between the response and the predictors under consideration) and 1 (perfect predictability by the model), although it is usually re-expressed as a percentage: it is more appealing to say, $R^2 = 76.5\%$ than to say $R^2 = 0.765$. The extremes never occur in realistic modeling problems. If you accidentally regress Y against itself, you will get 100%.

Adjusted R^2. Most statistics packages also report the so-called "adjusted R^2. This statistic shrinks R^2 to "penalize" it to account for the number of parameters in the model relative to the sample size. It is meant as a warning against "over-fitting"; in the context of simple linear regression, you could take it as a warning that the sample size is too small to yield trust-worthy results. In our experience, this does not occur very often; most researchers we have worked with have decently large samples. Rarely, when R^2 is very low, the adjusted version can even be negative. The computer doesn't care about what makes sense: it just does the calculation.

Sometimes a cartoon, or caricature, is helpful. Imagine having only two data points, yielding two dots on a scatterplot. This is over-fitting, taken to a ridiculous extreme. You could join the two perfectly with a straight line and then declare that your line explains *all* the variation in *Y*, and so R^2 would be 100%. This is also about 100% bogus. For the fun of it, we tried a small dataset with only three values; R^2 was 68%, while R^2_{adj} was 37%, a precipitous drop. Clearly, $n = 3$ is likely not a good idea. How big a drop should concern you? It is a judgment call, but we find a drop of 1% or 2% to be ignorable; a drop of 10% would get our attention. Some researchers report R^2 because of its interpretation; others favor R^2_{adj}. If they are quite similar in value, the interpretability of R^2 is appealing.

7.3 The ANOVA Table

All statistics packages report an ANOVA table to you. Here it is (Table 7.3) for the geyser data.

TABLE 7.3

ANOVA Table for the Geyser Data

Source	Df	Adj SS	Adj MS	F-Value	p-Value
Regression	1	40,644	40,644	1162	0.000
Error	270	9433	35	---	---
Total	271	50,087	---	---	---

Notes: "SS" stands for sums of squares, and "MS" for Mean square. The total SS is in fact the numerator in the variance calculation for the response variable. It is called the "adjusted SS" because each response value has been "adjusted" by subtracting the sample mean. The mean squares are calculated by dividing the relevant SS by its associated degrees of freedom. Then the ratio of the model MS to the error MS yields the *F* statistic, which is used to test whether the slope is meaningfully different from zero.

You don't need to worry too much about the ANOVA table, especially for simple linear regression. Back in the day, it was necessary to know how to calculate the numbers in an ANOVA table, so that you could then extract the useful bits.

For example, R^2 is SS (regression)/SS (total), then rescaled to a percentage. Also, the F statistic is simply the square of the t-statistics for the slope (see Table 7.1). We won't show you how to do those calculations; so long as you understand how to use the relevant summary values, you should be in good shape.

A usual summary value is the SD among the residuals (here $s_{res} = 5.91$; Table 7.2). As the last step in a step-by-step process, the SD is the square root of the relevant variance formula. The variance of the residuals is in Table 7.3 (Mean Square Error; 35 here). You won't often use this number directly unless, for instance, you need it to calculate the SE of a prediction, more on which below.

The *t*-statistic, calculated[3] as $t = (b_1 - 0)/\widehat{SE}(b_1) = 10.73/0.315 = 34.01$, has a couple of minor advantages over the F. For one thing, it will be positive or negative according to whether the slope coefficient is positive or negative. By squaring the t to get the F, the sense of direction is lost. Second, it has a direct interpretation. We can say that the observed slope (10.73) is about 34 standard errors away from zero, quite improbably far if the true slope is indeed zero.

7.4 Hypothesis Tests for Simple Linear Regression

Hypothesis testing is such a reflexive part of what we learn to do as practicing statisticians that we do it (and advisors, editors, and reviewers might insist on it) even when it might not be of much interest. Consider a data set with the length and weight of brown trout; such data might be collected by a management agency when studying the health of a brown trout population. Of COURSE longer trout will on average be heavier. Doing a test for that and showing a very small *p*-value tells us nothing we did not already know.

Table 7.1 shows typical hypothesis testing output for simple linear regression from a statistics package. Here, both the intercept and slope are highly significant (both *p*-values are very small), meaning we have strong evidence that the true slope and intercept are not zero.

Notes and comments:

Naming convention. Some statistics packages use the word "Constant" instead of "Intercept." Why? As we noted above, our usual intercept and slope naming of a regression equation is a convention, but not the only way it could be done. The regression line goes through the point $(\bar{Y}, \bar{X})$, so another naming convention could be $(\bar{Y}, \text{slope})$, in which case $\bar{Y}$ would be the constant in the equation. **Another one.** Most statistics packages use the name of the predictor (duration, here) rather than "slope." Why? If you have four predictors, you want to keep track of which slope coefficient goes with which predictor. So, they name the predictor. Fair enough.

SE estimates the SD of the relevant distribution. You understand, by now, that, for instance, the slope coefficient has a distribution in the sense of answering the question, "What might I get for numerical values if I were to repeat the study a very large number of times in the blink of an eye?" The SE of the coefficient is an estimate of the SD of that imagined distribution.

The numbers in the **t-value** column show how far from zero each coefficient is, in terms of the SE. For instance, the observed slope (10.73) is determined to be 34 SEs above 0, that being the relevant value if there was no linear relationship.

- By labeling the column with *t*, we are reminded that the resulting test of significance is done using the *t*-distribution.

- The value of 10.73 for the slope is an artifact of our choice of measurement units. Had we chosen to measure duration in hours instead of minutes, it would have been 60 times bigger (643.8) because then a one-unit change in time would be 60 times bigger. Had we chosen seconds, it would have been

0.18. When testing the significance of the slope coefficient, we are asking, "Is it close to zero?" Since our observed value is an artifact of our choice of measurement units, we shouldn't try to answer using our chosen scale. Measuring it in terms of SEs is more reasonable.

- Note on the previous note: **choose your measurement units carefully.** That choice won't change the relationship, but it *will* affect your ability to tell the story. For example, we routinely measure elevation in feet. A one-foot change in elevation on a mountainside is imperceptible, and so the slope for a relationship between tree height and elevation measured in feet would be a very tiny number. Perhaps a hectameter (a 100-meter increment would be a more compelling scale for telling the story in this case.

The regression is significant. The *p*-value for the slope is very tiny, but is not perfectly zero. It *is* zero down to many decimals, and so reporting it to be < 0.001 or even < 0.0001 usually does the job.

So is the intercept. Statistics packages routinely include a test of whether the intercept is zero. We suggest that you mostly just ignore that. Why? We ask you, "why would you care about the average value of Y when $X = 0$?" First, whether the average value of Y is zero for *any* particular value of X is likely not of any interest. Second, $X = 0$ is usually outside the range of interest for predictor values anyway. "Whisper words of wisdom, let it be"[4].

A one-tailed test of the slope. Suppose your research hypothesis regarding the relationship has a sense of direction (e.g., "we hypothesize the slope will be positive.") In that case, you are justified in doing a one-tailed test for the slope coefficient. Here's how you do it.

- If the estimated slope agrees with the direction of your hypothesis, then you may divide the reported *p*-value in half[5].
- If the estimate disagrees with the direction, however slightly, stop. The test is over. There is clearly no evidence in the data in favor of your research hypothesis. Indeed, the *p*-value will for certain be larger than 0.5.

7.5 Explaining the Effect of the Predictor

From Table 7.1, we can simply and easily state that a one-minute increase in the duration of an eruption is associated with a 10.7-minute increase in the waiting time, on average.

It is not enough, in a scientific report, to state the estimate (10.7) and stop there. Some statement of precision is called for. Reporting precision could take several forms:

Report a confidence interval. The usual choice for confidence level is 95%, so we will use that here. Find the appropriate t-multiplier from a t-table. To do so, you need the appropriate degrees of freedom, which are easy to determine. It is sample size (272 here) minus the number of parameters in the model (two here: the intercept and the slope). Using 270 degrees of freedom and selecting the value that captures the central 95% of the distribution, we get 1.97.

Then the CI is calculated as $b_1 \pm t_{df,CL} \times SE(b_1) = 10.73 \pm 1.97 \times 0.315 = (10.11, 11.35)$, which is valid assuming the distribution of the slope is approximately Normal. We are 95% sure the true slope is between 10.11 and 11.35. Given that most people use 95% as their confidence level and that for all but very small sample sizes, the t-multiplier is approximately 2, one could form an approximate 95% CI as $10.73 \pm 2 \times 0.315 = (10.1, 11.4)$. We are pretty darn sure the true slope is between 10.1 and 11.4.

It suffices (and is done by many writers) to simply report the estimate and its SE: (10.73, 0.315).

Aside: a note on causality.

In the above conclusion, we said "is associated with," not "causes." Causality can only be inferred in appropriately designed experiments. For instance, suppose the growth of a certain crop was being studied by growing seeds in small growth chambers. If several growth chambers are each assigned to be held at a certain temperature, then one could infer that changes in temperature are causally related to changes in growth rates.

7.6 Making Predictions from a Regression Model

In scientific research, explanations are often of more interest than predictions. Predictions are more often of interest in management settings; engineers might be very interested in predicting the strength of concrete being used to build a bridge, for instance. The geyser waiting times problem is arguably a prediction problem.

For a chosen value of the predictor, you might be interested in predicting the mean response, or you might be interested in predicting an individual response. The estimate will be the same, but the resulting intervals will not be. In both cases, you might write, "I am 95% confident that...," and so you might think the phrase "confidence interval" could be used for either. However, the convention has been adopted to call the interval estimate for the mean a CI, and for an individual value a prediction interval. This is a semantic choice.

It is typical that estimates are more precise near the predictor values and less precise as you get further from that point. There is less information feeding into the estimates at the extremes. That effect is subtle in our example because of the large sample size and that the values are spread out widely. In a usual dataset, there will be fewer values near the extremes, and the intervals will be quite noticeably wider there, as illustrated in Figure 7.2.

The 95% intervals are all formed by the fitted value of the regression model at X_0 plus and minus approximately two times the SE. However, the SE for predicting the mean of Y is smaller than that for predicting an individual value. You need to account for the precision of the mean *and* individual variation.

Luckily, many statistics packages will do this estimation for you. You simply specify a value for X, then it will generate a CI and PI for you.

7.6.1 Extrapolation

Extrapolation involves making predictions from the model for values of X that are beyond the range of observed values. In most statistics books and classes that discuss regression, the dangers of that are explained. In our collective experience, which includes thousands of consultations over the years with scientists from many disciplines, not once have we had occasion to say, "Actually, that's a bad idea. Let me explain," and then go on to warn about extrapolation.

Why is this? There are several possible explanations.

1. We are such brilliant teachers that students never forget any lessons we impart to them. We strike fear into their hearts, and they never consider extrapolating ever again.

2. Scientists are usually more interested in explanation than prediction. "How do changes in X affect Y?"

3. When prediction is of interest, their research studies are built around predictor values in the range of interest.

One of these explanations is highly implausible.

The following example (Figure 7.2) shows a regression model for a small data set. The smallness is intentional, as it dramatizes a certain feature of making predictions. Live weight and so-called dressed or slaughter weight were recorded for nine small beef cattle. Included in the figure are the regression line and a band showing a so-called 95% confidence band.

Every regression goes through the point $(\bar{Y}, \bar{X})$, indicated by the asterisk in Figure 7.2. You are most sure of the location of the regression line at $\bar{X}$; the confidence band is at its narrowest there, and it gets wider as you move away from that central point. So, if you try to predict beyond the range of your data, your predictions will become less and less precise. Furthermore, you need

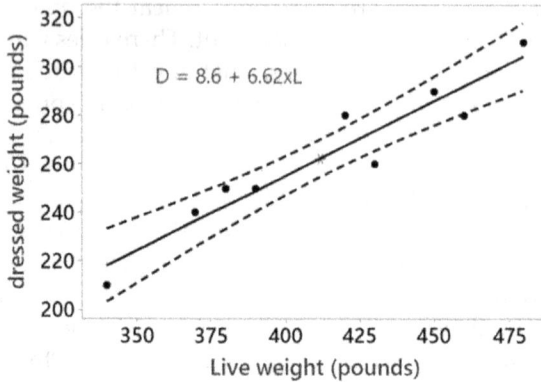

FIGURE 7.2
Regression of dressed weight on live weight, with a 95% confidence band.

to be able to argue that the model continues to be appropriate beyond the observed range. In this case, you might not worry too much if you want to make a prediction for, say, a 490-pound animal. But notice that the model predicts a dressed weight of 8.6 pounds for an animal whose live weight is zero!

7.7 Chapter Summary

We examined the numerical output usually associated with a regression model, and explained the meaning and various uses for that output. We examined

- The regression coefficients, and their use in estimation and testing,
- The coefficient of determination (R^2) and its cousin (R^2_{adj}),
- The ANOVA table.

We went into some detail on testing and estimation, including prediction. For the most part, you can count on your statistics package to do the work. It is important for you to know how to use the results of that work.

Notes

1 ANOVA is an acronym for "Analysis of Variance." Just about every statistical analysis studies the variation in a data set, attempting to pin down how much of that variation is due to a model, and how much is left over as "random." You could call just about any analysis an ANOVA, but we don't. Cultural norms...

2 You might think this is "pre-historic," but Ken remembers drawing graphs this way. Don't think about that too long...

3 The formula is usually written this way to remind us that we subtract the value under the null hypothesis from the statistic. Here, of course, that value is 0.

4 With apologies to Paul McCartney.

5 This is a benefit of a one-tailed test. A smaller p-value leaves you more likely to declare a significant finding, i.e. the test has more power than the analogous two-tailed test. The price, though, is that you have to stick your neck out and, *a priori*, declare a direction. Good judgment is called for.

8

Background Reading and a Few New Ideas

Most of this chapter qualifies as "background reading" for those interested. It is not directly instructional, but might make some of the ideas around regression less opaque. This includes a discussion of the origins of the conventional formulation of a regression line and another peek into history to learn how the word "regression" came to be applied to these models. Also, the reason it is a linear model is probably not what you think it is.

Some practitioners use a graphical method called a Normal Probability Plot to examine the Normality assumption of the residuals. We don't find it as intuitive to use as a histogram, but it is indeed a legitimate tool, and some statistics packages provide a p-value to test that Normality.

It sometimes occurs to a practitioner that it is easy algebra to rewrite the regression equation so that it predicts X given Y. We discuss that and show why it is not a good idea. The punch line? Rerun the regression after swapping the roles of the response and predictor variables.

What is novel? There is a "bonus section" that introduces an interesting and conceptually simple regression method for a setting which violates the usual assumptions, in particular equal scatter. It is called ratio regression and is particularly applicable when values of the response variable are proportional to those of the predictor. We are not quite sure why it is not routinely taught, but maybe this section will rectify that a little.

And for those who want to dive deeper, we describe in the appendix the math behind SLR.

8.1 On the Names "Regression" and "Linear Model"

Why is this method called "regression"? The term is somewhat opaque. Francis Galton, a cousin of Charles Darwin, was a British polymath active in the 19th century; statistics was one of his many interests. The canonical data set, often discussed in the context of the origin of the term regression, is one where he studied the heights of fathers as the predictor, and the heights of their sons as the response. We will discuss those data and then the geyser data in a moment, but let's start with a thought experiment[1].

Consider a group of imaginary students taking a test composed of 100 true/false questions, who choose all their answers randomly. Repeat for a

DOI: 10.1201/9781003609605-10

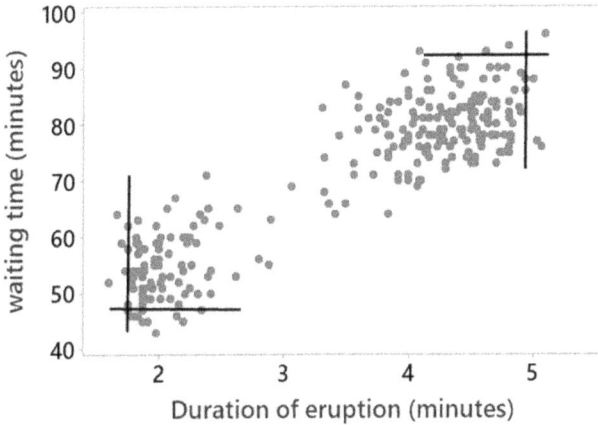

FIGURE 8.1
Geyser waiting times and durations with extremes marked.

second test. It should appeal to intuition that the "students" with the lowest scores in test one will randomly be closer to the mean in test two, ditto for the highest scores. Regression to the mean in action. In real life, test scores will come from a combination of luck and skill, so the two sets of test scores will be correlated. It is a feature of the behavior of correlated data, as shown in Figure 8.1, using the geyser data as an example.

The very extremes of duration values (demarked by vertical lines) are not associated with the very extremes of waiting times (horizontal lines) and vice versa. In each case, there is a tendency for the values of the other variable to be closer to their respective mean—that is, to be not as extreme. Galton referred to this as "regression to the mean."[2] There are a few more details, but this led Galton to refer to the estimation method as "regression." If the technique were to have been invented today, it might have become a "simple linear model" rather than a "simple linear regression." And then you would not have had to read this explanation.

8.1.1 While We're at It: Why is It Called a "Linear Model"?

Fair warning: this section gets a little nerdy. If you are new to this business, we suppose you think the answer is obvious: "Because the model is a straight line." That is not so; it is a coincidence. Most linear models do not yield straight line fits. Hunh? The name comes from the mathematical concept of a "linear combination of the parameters."

A simple linear regression is composed of the sum of β_0 and $\beta_1 \times X$. In a more complicated model, we might have more than one predictor, so let's consider $\beta_2 \times X_2$, and relabel the foregoing as $\beta_1 \times X_1$. And we might have one of them squared; let's use $\beta_3 \times X_2^2$. So the entire model could look like $\beta_0 + \beta_1 X_1 + \beta_2 X_2 + \beta_3 \times X_2^2$.

And of course, we needn't stop at four terms, nor at only two predictors. The point of all this is that the model is an example of a structure that mathematicians call a "linear combination of the parameters," and hence models that fit into this form are called linear models. Note that a model with X and X^2 would yield a curved line, not a straight line. It is only in the case of simple linear regression that the model is indeed a straight line. As we said at the outset, this is a mere coincidence.

8.2 The Normal Probability Plot and Reverse-Engineering a Regression Equation

Here, we mention an alternative to the histogram of the residuals as a tool for checking whether their distribution is approximately Normal. Also, we examine the utility of reverse-engineering a regression equation so as to use it to predict X from Y.

The Normal probability plot. A Normal Probability Plot can be used to assess the Normality of residuals; we don't prefer it; that said, we will show you one (Figure 8.2) for the geyser data and then discuss why we prefer a simple histogram.

This plot was invented when drawing a histogram was a tedious pencil and paper task. A calculation is done on the residuals that have the property that, if they come from a Normal distribution, they would fall on a straight line in this plot. They do not here; if you learn to use this tool, you will learn that the dots falling below the line at the left side of the graph imply the distribution is skewed right. Fine, and using it is not wrong, but we claim that

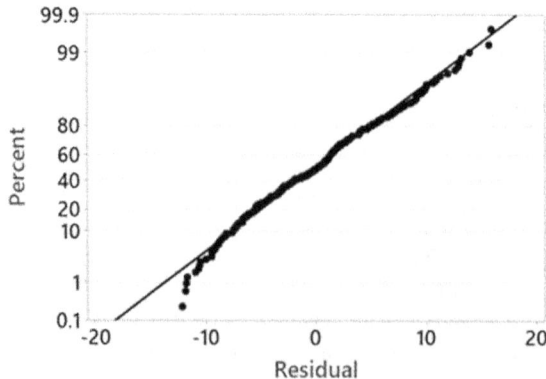

FIGURE 8.2
Normal probability plot for the residuals from the geyser data.

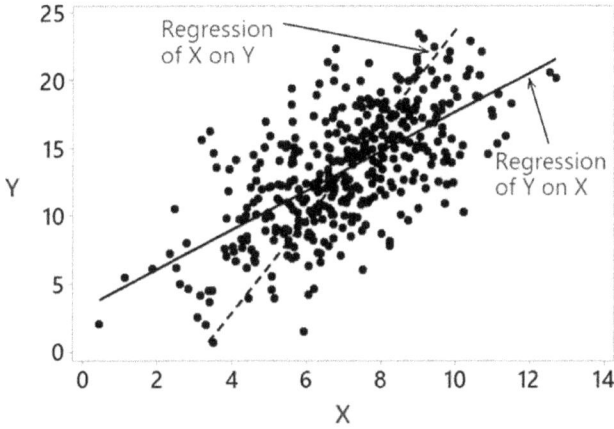

FIGURE 8.3
A demonstration of the fallacy of algebraically inverting a regression line.

a quick glance at the histogram of the residuals will lead you to that same conclusion in a more intuitive and direct way.

8.2.1 Predicting *X* from *Y* Using the Regression of *Y* on *X*

Once one has regressed *Y* on *X*, obtaining an estimated line in the form of $Y = b_0 + b_1 X$, it is sorely tempting to use this equation to find *X* given *Y*; this is a simple algebra problem: $X = \left(\frac{-b_0}{b_1}\right) + \left(\frac{1}{b_1}\right) Y$. Unfortunately, this will likely yield biased estimates of the average of *X* given *Y*. The solid line in Figure 8.3 is the regression of *Y* on X. The purpose of a regression line is to estimate the *average* of the response given a value of the predictor. The solid line in Figure 8.3 splits the data evenly above and below the line for given values of *X*. For instance, given *X* = 4, the line shows that the average of *Y* is approximately 8. Now pick, for instance, *Y* = 20, and notice that the line does not do a good job of finding the mean of *X* for that value. We say it produces biased estimates. The dashed line in Figure 8.3 splits the data horizontally and is in fact the regression line of *X* on *Y* in this figure. To get this, we ran the regression with the roles of the response and predictor reversed.

8.3 Ratio Regression—the Forgotten Model

Consider the following data. The goal is to estimate the density of prey for the Amur[3] tiger; their prey base consists of several species of deer and wild boar; all are included in this illustration. The response variable is the number

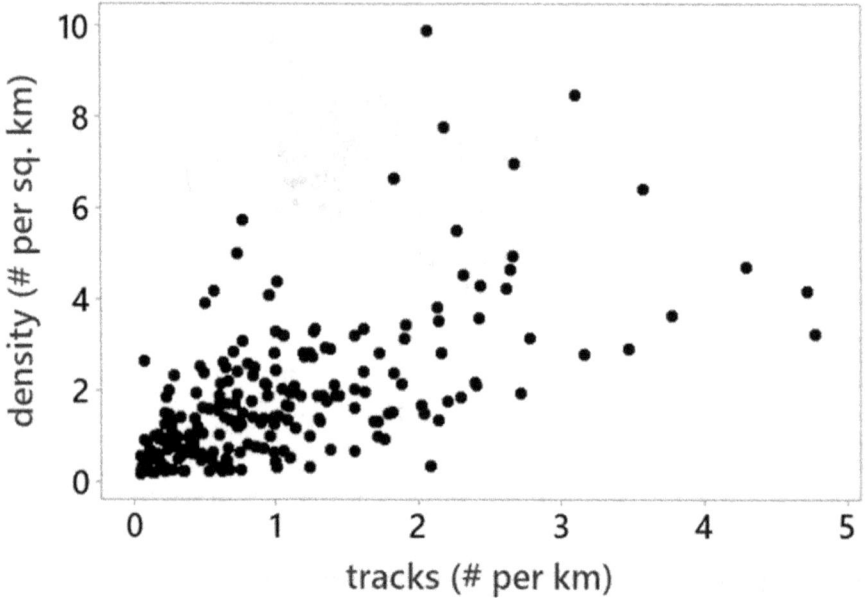

FIGURE 8.4
Observed prey density plotted against track counts for Amur tiger prey.

of animals per square kilometer; the predictor is the number of tracks per km counted along the perimeter of a 10 km × 10 km square. In the study, two biologists traversed the perimeter, counting tracks shortly after a snowstorm. The snow would have covered old tracks, so any tracks seen were sure to be fresh. Of course, some of the counted tracks might well have been from the same animal more than once. Then a team of people walked through the plot, being loud and clearly present. There were others waiting outside the plot to count the number of animals that fled. If there is a reasonable relationship, ecologists could use the track-count method to estimate prey abundance in a manner that is relatively easy to carry out and that has little impact on the animals themselves. Figure 8.4 shows a scatter plot of the data.

The problem, of course, is the heteroscedasticity. When Ken first encountered these data, it was clear how to proceed. He used a logarithmic transformation on both variables and fitted a simple linear regression on that scale (Figure 8.5).

This is not perfect, but this is arguably a better fit to the usual regression model assumptions than the untransformed version. Ken came across these data while working with ecologists at the Wildlife Institute of India, where he visited while on sabbatical at the University of Pune, India, in 2002. A couple of years later, he was invited by George Casella to contribute an essay to *Statistics: A Guide to the Unknown*, a book that had come out with a new edition every ten years or so. Each edition had a series of new essays highlighting the role of statistics in the real world. He was delighted to be invited to contribute, and these

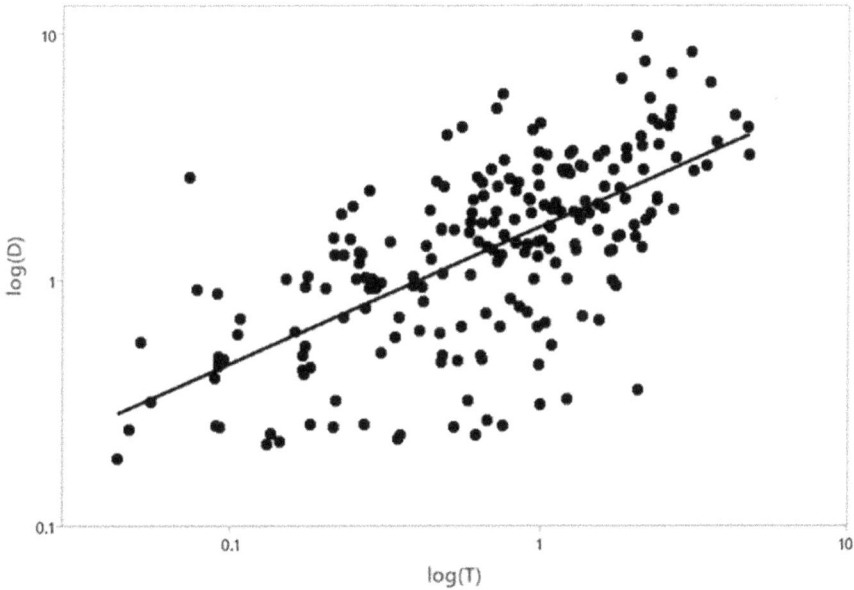

FIGURE 8.5
Log-transformed response and predictor values, with the fitted regression line. The equation for the line is $\log(D) = 0.22 + 0.55 \times \log(T)$.

animal abundance and track count data came to mind. That led to the log-log modeling appearing in the fourth edition of Gerow *et al* (2006).

This is a Tyler Johnson moment (remember Tyler?).

These data were intended to be used to predict the number of animals in such a 10 km × 10 km plot, following a suitable perimeter track count. The regression model does indeed predict the average of log(Y) given log(X), but that's not what we want in the end. The mean of log(Y) back-transforms to the median of Y, not the mean (Ramsey and Shafer, 2002, section 8.4). There is a correction that can be made to yield an estimate of the mean, which depends on the data having a so-called log-Normal distribution.

Fast forward a decade or so... Ken is teaching sampling methods and teaches a sampling method called ratio estimation. At the heart of it is a regression model called ratio regression. It is well-known and oft used in the sampling literature (e.g., Scheaffer et al, 2006, section 6.3; and Lohr, 1999, section 3.4); somewhat surprisingly, it is quite less well known and hence not used outside of that literature. We think it deserves a higher profile. We'll demonstrate it first and then show the model details. Figure 8.6 illustrates the ratio regression and the back-transformed log-log regression for these data.

These data reasonably have the property that the response (density of animals, Y) is proportional to the predictor (track counts, X); in symbols[4] $Y \propto X$. The ratio regression model is of the form $\mu_{Y|X} = \beta X$, where the model has no

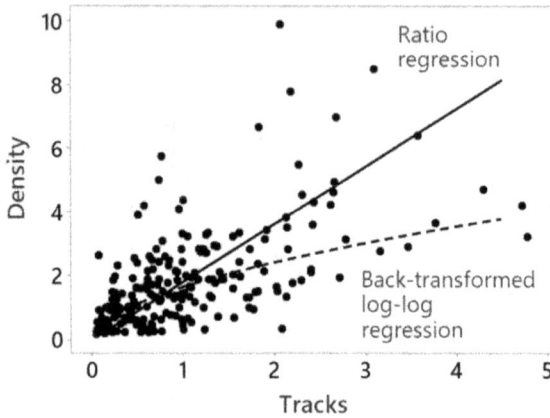

FIGURE 8.6
Ratio regression and back-transformed log-log regression models for the prey data.

intercept and the slope β rescales to get from X to Y. It is also assumed that $\sigma^2_{Y|X} = X\sigma^2$; in other words, the variance of Y is assumed to be proportional to X. The method is relatively robust to that assumption; so long as variation somehow grows with X, things are often fine.

With these assumptions, the best linear estimator of β is

$$\hat{\beta} = \frac{\sum Y_i}{\sum X_i} = \frac{\sum Y_i/n}{\sum X_i/n} = \frac{\bar{Y}}{\bar{X}}.$$

The estimator is shown in two forms. The first is the ratio of the totals of the response and the predictor; the second is the ratio of their means. The estimator is commonly called a ratio of means estimator. The statistic is intuitive to interpret, but there are complications in using it

Here, the numerator and denominator are both random variables, and they are correlated with each other. This leads to complications in estimating the variance of $\hat{\beta}$ (see below), and the Central Limit Theorem does not apply to the ratio of two means, so we recommend bootstrapping for inference with ratio regression. In fact, we introduce these details in Chapter 15, where the ratio of means $\frac{\bar{Y}}{\bar{X}}$ appears in the context of making inference on relative change in means from paired and two independent samples data.

8.4 Chapter Summary

In some sense, this is a bonus chapter, giving you some background reading into regression and, for those who want it, a deeper dive into the math behind simple linear regression.

We discussed the origins of the conventional formulation of a regression line and took a peek into history to learn where the word "regression" came from. Also, we shared the secret behind the name, "linear model." You are now officially one of the cognoscenti.

We took a moment to discuss the Normal Probability Plot, which some folks use to examine the Normality assumption of the residuals. We don't find it as intuitive to use as a histogram, but it is indeed a legitimate tool. We also demonstrate the futility of reverse-engineering the regression model, should you, in fact, want to predict X from Y. (You need to run the regression model with their roles reversed.

In addition, we introduced you to an interesting, somewhat unknown regression model called ratio regression, suitable when values of the response variable are proportional to those of the predictor.

Appendix: The Math behind a Simple Linear Regression

A.1 Modeling Details and Formulae for Slope and Intercept

Alert: If examining mathematical formulae works for you, a deeper understanding can come from this section. If that is not so, you can skip this section and likely not have a worse life for it.

With the (intercept, slope) formulation in mind, here is a conventional way to present the model as an abstraction:

$$Y = \beta_0 + \beta_1 X + \varepsilon;$$

where

1. Y symbolizes the response variable, X the predictor.
2. The Greek letter β (beta) symbolizes[5] the intercept (subscripted by 0) and slope (subscript 1)[6], and
3. Epsilon (ε) represents[7] "random error".

To get *really* detailed and proper about this, many authors note that we assume that the random error has an approximate Normal distribution, with variance σ^2, commonly abbreviated into symbols as $\varepsilon \sim N(0,\sigma^2)$. That whole thing gets unpacked as the random variation is assumed to have a Normal distribution with mean 0 and variance σ^2.

Notes:

1. We need not assume that the random variation has a Normal distribution. We *do* need that the slope and intercept estimates have at

least approximately a Normal distribution, since that assumption drives *p*-value calculations, confidence intervals, and so on. But that Normality can be purchased by having a large enough sample size (the Central Limit Theorem in action; see Chapter 2). So, we can relax that bit to the question, "Do the residuals have a Normal enough distribution?"

2. Sure, there has to be a single variance (the homoscedasticity assumption); without that, the standard error formulae are invalid, for instance. But *Y* comes from some population with a variance, and so does X. Maybe another subscript would be a good idea (σ_R^2 for instance).

The estimated model can be written as $\hat{Y}_X = b_0 + b_1 X$, where b_0 and b_1 are the estimated values of the intercept and slope. The "^" (carat, "hat") on *Y* is to remind us that it is an estimate, and the subscript *X* reminds us that the estimate of the mean of *Y* depends on *X*.

Given an appropriate sample, the slope is estimated by $b_1 := \frac{r s_Y}{s_X}$, where *r* is the observed correlation (more on which below) between *Y* and *X*, and S_Y S_X are the sample SDs for *Y* and *X*. Then the intercept is estimated by $b_0 = \bar{Y} - b_1\bar{X}$. The sample-based slope and the intercept are random variables, meaning they vary among repeats of the study. Given a sufficient sample size, we can usually assume they have an approximate Normal distribution. We still need to estimate their respective standard errors.

A.2 Standard Error Formulae

The standard error of a statistic is an estimate of the standard deviation in the distribution of the statistic. Some authors currently use SD, some SE. We will use both interchangeably. Given sample values and estimates for the parameters,

First, $\widehat{SE}(b_1) = s_r \sqrt{\frac{1}{(n-1)s_x^2}}$, and $\widehat{SE}(b_0) = s_r \sqrt{\frac{1}{n} + \frac{\bar{x}^2}{(n-1)s_x^2}}$, where

- s_r is the SD among the residuals,
- s_x^2 is the variance of the sample of *x* values,
- *n* is the sample size, and
- $\bar{x}$ is the sample mean of the predictor values.

Behavioral notes:

- The more random variation in the residuals (i.e., larger s_r), the harder it will be to estimate the line tightly (i.e., the *SE*s get larger)
- Having predictor values that are more spread out (larger s_x^2) leads to sharper estimates, as does larger sample sizes.

Notes

1 This conceptual exercise was drawn from the Wikipedia: Regression toward the mean - Wikipedia Go there if you want their more detailed discussion.

2 If you think about this for heights of parents and their children, you might feel grateful that this occurs. Suppose it did not. Eventually we would have biologically untenably short people and untenably tall...

3 Commonly called the Siberian tiger, it lives largely in the Amur river basin in Russia's northeast.

4 There is a slight bit of misbehavior here when $X = 0$. Strictly speaking, the model suggests that if $X = 0$, then so does Y. Here, if you observe no tracks, there might still be a few animals in the 100 square kilometer plot). So asserting the $Y = 0$ when $X = 0$ is off a bit. Tyler would be OK with this.

5 This formulation is pretty universal in statistics texts; you might have seen it as "mX + b," with "m" being the slope and "b" the intercept, in other places.

6 One wonders why the coefficient labeling starts with 0 instead of 1. Yep; we wonder too...

7 The term "error" makes us cringe slightly, since there is no "error" has been made. It is just a reminder that, on average, Y might be reasonably described as having a straight-line relationship with X, but there will still be some additional random variation present.

References

Gerow, K. D. Miquelle, and V. V. Aramile. 2006. Monitoring Tiger Prey Abundance in the Russian Far East. In R. Peck, Ed., "Statistics, a Guide to the Unknown". Thomson/Duxbury, Belmont CA. 440 pages.

Lohr, S. 1999. Sampling: Design and Analysis. Duxbury Press, Pacific Grove, CA. 494 pages.

Ramsey, F., and D. Shafer. 2002. The Statistical Sleuth, (2nd Edition). Duxbury Press, Pacific Grove, CA. 742 pages.

Scheaffer, D., W. Mendenhall, and L. Ott. 2006. Elementary Survey Sampling, (6th Edition. Duxbury Press, Pacific Grove, CA. 464 pages.

9

The Use of Logarithms in Regression Models

What's novel? We think the whole chapter is, since folks commonly use logarithmic transformations because their data is skewed. That rationale smacks slightly of statistical voodoo and doesn't lend itself to any story-telling intuition. As it happens, the relationship between the response and the predictors is often relative. This relativity might exist for the response variable. For example, in a setting of exponential increase, for a specified increase in X, there is some percentage increase in Y. This relativity might exist for the predictor or, often, both.

Relativity is very common in biological and other relationships, and logarithmic transformations capture that relativity. We would even say it does so beautifully. This chapter is devoted to revealing and explaining this role.

Most scientific research is aimed at explaining the effect of this or that predictor or factor. In regression models, that focus requires an interpretation of slope coefficients. Unfortunately, when folks use logarithmic transformations, they get stymied by the math and stop after reporting p-values. So, we end the chapter with a section that will show you how to do the math.

9.1 Relativity in the Response

As we go through this chapter, we will refrain from showing you all the output for the various models we consider. Instead, we will focus on illustrating their properties as they pertain to the topic of this chapter, namely relativity.

Let's begin by revisiting an example from Chapter 6. Huber *et al* (2024) studied levels of regulatory and functional adrenal gland proteins across the life course, using baboons as a model. Citrate synthase (CS, measured by visible presence (% area) in stained adrenal gland cells), a measure of mitochondrial function, was one of their response variables. We saw in Section 6.4 that the regression fit to the data was much improved by log-transforming the CS values. Figure 6.15a shows exponential decay, back transformed after using a logarithmic transformation on CS, repeated here in Figure 9.1.

The fitted line is $\log(CS) = 2.47 - 0.093A$, which, on the original scale is $CS = 10^{2.5} \times 10^{-0.095A}$. Recall that addition (subtraction) on the logarithmic scale translates to multiplication (division) when back transformed. Multiplication

DOI: 10.1201/9781003609605-11

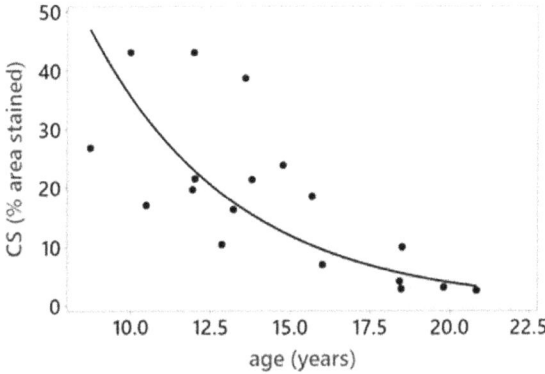

FIGURE 9.1
Regression relationship for citrate synthase with age.

on the log-scale becomes exponentiation when back transformed, which is why the predictor A is now in the exponent. The first form is a classical SLR model. If you subtract the estimate for some level A of age (let's call it $\log(CS)_A$) from the value at $A+1$, $(\log(CS)_{A+1})$, you will see that everything cancels out except the slope: -0.093. Hence, our usual interpretation for the slope applies: it is the change in the average of $\log(CS)$ for a one-year increase in age. A simple interpretation, to be sure, but the response variable is not intuitive to understand.

If you try that subtraction for the back-transformed version, it will just stare at you: $CS_{A+1} - CS_A = 10^{2.47} \times 10^{-0.095(A+1)} - 10^{2.47} \times 10^{-0.095(A)}$. Nothing cancels. However, if you divide instead of subtracting, everything cancels out except the piece that includes the slope $(10^{-0.095})$: $\frac{CS_{A+1}}{CS_A} = \frac{10^{2.47} \times 10^{-0.095(A+1)}}{10^{2.47} \times 10^{-0.095A}} = \frac{10^{2.47} \times 10^{-0.095A} \times 10^{-0.095}}{10^{2.47} \times 10^{-0.095A}} = 10^{-0.095} = 0.80$. For multiplicative change, the "balancing point" (i.e. no change) is one, not zero. Here you could say, "CS values are 80% of what they were," but that is awkward, even when rescaling to a percentage. When the answer is less than one, it is easier to talk about the size of the drop. In this case, we would say that CS values drop by 20% for every advancing year of age $((1-0.80) \times 100\% = 20\%)$.

Where are we going with all this? Here are the four main ideas:

Many relationships between a response and a set of predictors are relative. Perhaps

- The response changes in relative ways to additive changes in a predictor, or
- There is an additive change in a response to multiplicative changes in a predictor, or
- Relative change in the response is associated with relative change in a predictor.

That relativity is expressed as curved relationships that can be treated with standard regression models if (order follows the list just above)

- The response is log-transformed,
- A predictor is log-transformed, or
- Both are log-transformed.

The assumption-checking can tell you if you have found the right scale. As you continue into this chapter, you will learn how to interpret those relationships, step by step.

Relativity is common in regression relationships. If you work in a certain field long enough, you know going into an analysis that you will have to log-transform Y, or X, or both. Fisheries ecologists, for example, will expect to log-transform both weight and length. If you don't know ahead of time, it is really a wonderful thing that doing the assumption-checking can guide you. For instance,

- Curvature and increased scatter in association with increased fitted values points towards relativity in Y (i.e. log-transform it).
- Curvature alone, but with scatter not changing might suggest[1] relativity in X (log-transform it).
- But be careful: curvature and increased scatter can show up when there is relativity in both Y and X.

We have just seen an example where the relativity lives in the response, accommodated by log-transforming it. Now we will work through several illustrative examples.

9.2 The Species-Area Relationship: Relativity in X

Newmark (1986) used data from national parks in the western United States to study the famous "species-area" relationship. The data (area in square kilometers and species richness[2]) are from twenty-three national parks. Each park is an "island" of suitable habitat, surrounded by less suitable habitat. From a conceptual point of view, islands in this species-area setting should be isolated so that the plants or animals of interest cannot easily get from one island to another. Four summary graphs from our examination of this relationship are shown in Figure 9.2.

There is clearly an increase in species richness with increased area, but the relationship is not well modeled by a straight-line regression (Figure 9.2a). In this case, the relationship is one where a certain number of species is

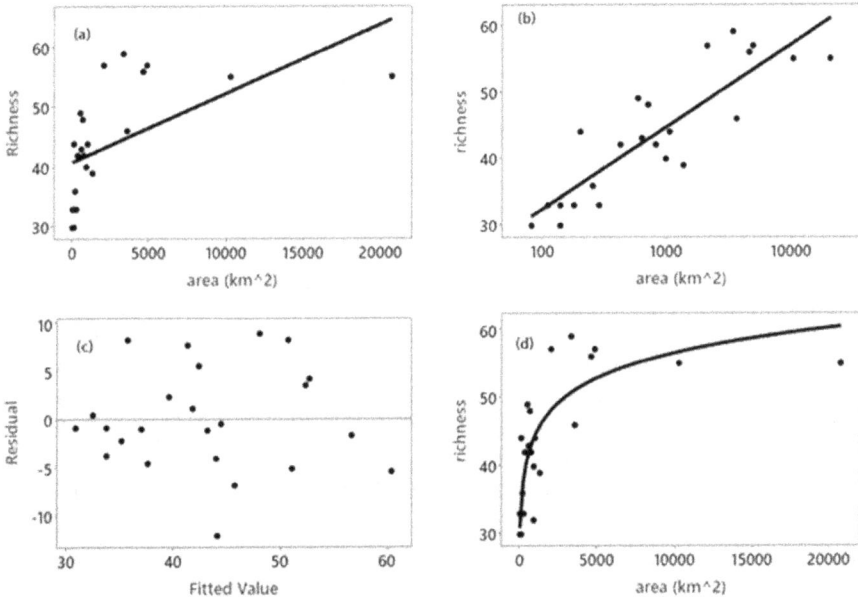

FIGURE 9.2
Four graphs of the species-area relationship for Newmark's data. (a) A regression applied to the original data. (b) The area has been $\log_{10}$-transformed. (c) The residuals versus fits graph for that model. (d) The fitted line on the original scale.

added for a multiplicative increase in area. This conclusion rests on the fact that a straight-line model fits well when the area has been log-transformed (Figure 9.2b). Assumptions of goodness of fit and equal scatter are well met (Figure 9.2c); the distribution of the residuals is reasonably "Normal enough" (not shown). Specifically, there is an increase in 12.3 species for every tenfold increase in area (Figure 9.2d).

It is useful to point out here that there is a quick and easy interpretation of the slope due to having used 10 as the base for the logarithmic transformation of area. For every tenfold increase in area, there are, on average, 12.4 more species. To speak of tenfold increases in area is not unreasonable: there are several tenfold increases from the smallest to the largest park. If you are inclined to use e (the base of the so-called natural logarithm) in your work, you would be mildly stuck at this point since it is not so natural to speak of 2.71828-fold increases in area.

Here is how to change the scale of inference to one of your choosing. For example, suppose you wanted to talk about the species-area relationship for Newmark's data in terms of doublings of area (twofold increases). It's simple: you multiply the fitted slope by your chosen scale using the base that was used in the modeling. Here we would get $12.83 \times \log_{10}(2) = 12.83 \times 0.301 = 3.86$. There is an almost fourfold increase in richness for every doubling of area.

Note that had you used the natural logarithm, the observed slope would be 5.77. If you apply the foregoing arithmetic to 5.77, but use e for the base instead of 10, you will get the same answer. The logarithmic base you choose does not change the story.

9.3 Morphometrics Example: Trout Weight-Length

The relationship between the length of fish of a given species and their weight is of great interest to fisheries scientists. Each species has its own characteristic shape. Much work has been done over the years in establishing the length-weight relationship to assess the health of populations of a species. Figure 9.3 shows such a relationship for cutthroat trout (*Oncorhynchus clarkii*) in a river system in the western United States (Dauwalter et al, 2022). In panel (a), the data are in their original form; in (b), weight has been log-transformed; in (c), length has been. Finally, in panel (d), both have been log-transformed[3].

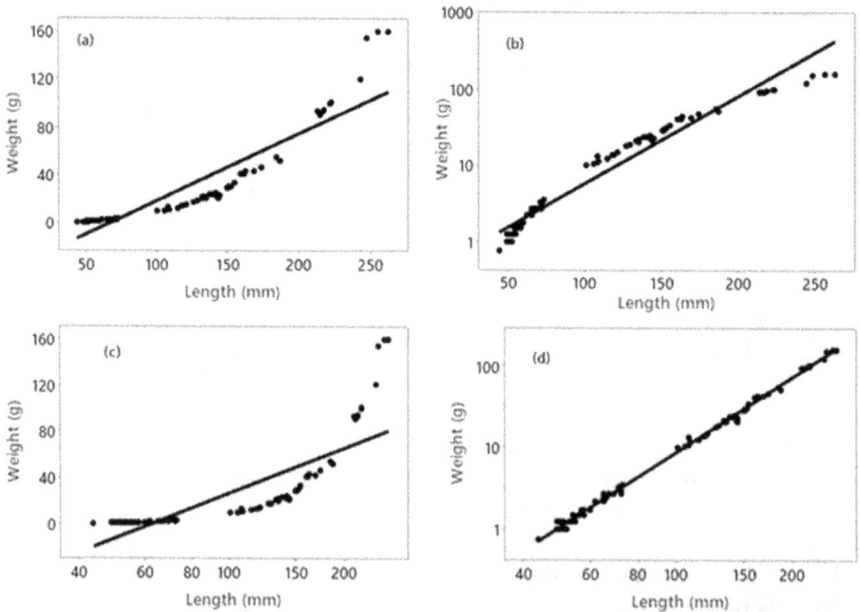

FIGURE 9.3
Simple linear regression models fitted to four versions of trout weight/length data. Both are in their original form in panel (a), weight has been log-transformed in panel (b), length has been transformed in panel (c), and both are transformed into panel (d).

Even a quick glance shows that only the relationship in panel (d) shows a good fit between a straight line and the data values. The others are all terrible. There is a theoretically based model for this relationship, but let's continue by way of exploration, and then see how our empirical approach lines up with that model. Let's study the log-log model further.[4] In particular, let's learn how to interpret the relationship: how do changes in length predict changes in weight, using the fitted model shown in panel (d). The fitted equation is $\log(W) = -5.14 + 3.04 \times \log(L)$. Here, we need to think in terms of relative change for both W and L.

First, let's restate the model on the original scale by raising both sides to a power of 10, because 10 was the chosen base for the logarithmic transformations. This gives us $W = 10^{-5.14} \times L^{3.04}$. Since the exponent of L is essentially "3", we could say that weight is a fraction of length cubed. As such, this fraction might be a characterizing feature of that species, which could then be compared across species.

We need to think in terms of relative change for length. A tenfold increase in length is not convenient for telling the story, given that there is only a fivefold change of length in the data. Instead, we could think of a doubling, or of a 50% or 10% increase. For simplicity, let's work in terms of doubling length. Following the nomenclature style we used for the CS example, $\frac{W_{2L}}{W_L} = \frac{10^{-5.14} \times (2L)^{3.04}}{10^{-5.14} \times L^{3.04}} = \frac{2^{3.04} \times L^{3.04}}{L^{3.04}} = 2^{3.04} \approx 8$. For every doubling in length, weight is about eight times greater. A visual inspection of Figure 9.3a shows the weights being just below 10 g for length 100 mm, and just below 80 g for 200 mm.

All that said, doubling length is a big jump. If we do the same arithmetic for a 10% increase, we will get $1.1^{3.04} \approx 1.34$: there is approximately a 33% increase in weight associated with a 10% increase in length. Similarly, $1.5^{3.04} \approx 3.4$: a 50% increase in length is associated with approximately a 3.4-fold increase in weight. The choice of the size of change in length is a story-telling choice.

As it turns out, the use of a log-log relationship for fish length/weight data has a theoretical underpinning. Consider a 20 cm long trout. Cube that length, yielding a cube of size 8000 cm³ (that sounds like a lot, but it might help to visualize the cube being about 8 inches on a side). Now, a given trout, having trout shape, will have a volume that is some small portion of that cube. And if we imagine an average density (number of grams per cm³) for a trout, then we might argue that an idealized relationship between length and weight might be $W = c \times L^3$, where c is a correction term scaling the cube to account for shape and average density.

In this model, relative changes in length are associated with relative changes in weight. This relationship takes on the form $\log(W) = \log(c) + 3 \times \log(L)$. Notice that in our fitted relationship, the slope was 3.04, quite close to 3. The value of three indicates that growth may be isometric, that is, the shape doesn't change as the fish grows. If it is less than 3, then a fish becomes less rotund as it increases in length; if greater, it becomes more rotund.

9.4 Choice of Scale, Choice of Base

The choice of measurement units can affect your ability to tell a statistical tale; it doesn't usually change the relationship, only the ability to discuss it reasonably. Sometimes, a standard choice of measurement units can get in the way of storytelling, and you should feel free to consider other choices. And you saw in the foregoing examples that the choice of base for relative change can similarly affect the quality of the story the data can tell. Such choices do not change the story (the relationship is the relationship is the relationship), only the ease of relaying it.

It does not really matter what base you use for logarithmic transformations. Some disciplines routinely use base 10, others use base *e*, the so-called natural logarithm. We recommend using the same base for the response and the predictor if you choose to transform both. That will keep the subsequent arithmetic much cleaner.

9.5 The Species-Area Relationship: It's All Relative

This is a Tyler Johnson moment (remember him?). As we saw in section 9.2, Newmark's species-area data are well modeled by log-transforming area: you add a certain number of species for a given relative increase in area. If you log-transform both species and area, the statistical qualities of the relationship are similar to having log-transformed area only. The assumptions appear equally well met. Indeed, the residuals versus fits graph is very, very similar. We will show you here the fitted model on the original scale (Figure 9.4). Compare it to Figure 9.2d. It is difficult to tell them apart.

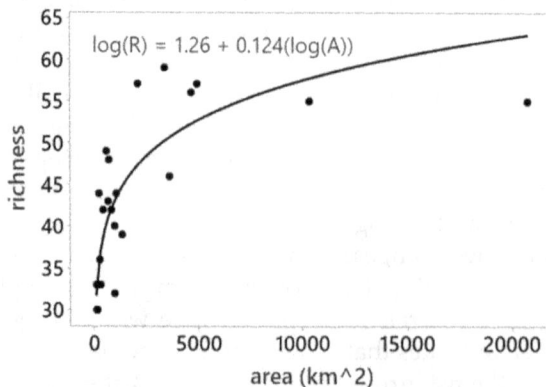

FIGURE 9.4
Newmark's species-area relationship with both variable log-transformed.

A tenfold increase in area is associated with a 33% increase in the number of mammal species: $10^{0.124} = 1.33$. Compare this to the inference from our first encounter, where there are approximately 12 more species for a tenfold increase in area. Further, R^2 is very similar for the two models (69% with only area transformed, and 68.3% for both transformed). What to make of this?

First, there are only approximately two tenfold increases from the smallest park to the largest, so there is not much opportunity for differences in predictions to blow up. Second, the number of species ranges from about 30 to almost 60, not even one doubling. Since there is not very much relative change in the number of species, log-transforming species does not have a dramatic effect. Which would you choose? Log-transform area only, or both? Long live Tyler Johnson!

9.6 Log-Transformations as Sleight of Hand (and Some Protective Medicine)

Often, the motivation for log-transformation goes something like the following. Given a curved relationship between Y and X, and heteroscedastic as well (e.g. Figure 9.1), log-transforming Y will often yield a statistical relationship that will be well modeled by a straight-line relationship. In examining the untransformed and transformed analyses, you might also notice that R^2 is higher in the log-transformed case. To many, this smacks of statistical sleight of hand: some slightly mysterious statistical magic being done to create the illusion of a nice relationship where none exists. In cases where the log transformation does indeed improve the statistical relationship, we contend that the reason is that it better reflects the biology, as we have illustrated in this chapter.

In regression models, the focus is often on an interpretation of slope coefficients. Unfortunately, when folks use logarithmic transformations, they get stymied by the math and stop after reporting p-values. So, we will show you here how to report those effects. Some of what is here is repeated from the sections above. We do so in order to put all the mathematical arcana into one spot for your convenience.

Before we begin, though, we need to explain that **when you have log-transformed the response variable, you need to discuss the effect of the predictor in terms of the *median* of Y,** not the mean. As the examples in this chapter have shown, a good choice of scale (log-Y, log-X, or both) will usually show that the statistical assumptions are well-enough met. If all is well, the implication is that the distribution of the regression coefficients is approximately Normal. In that case, so is the distribution of the mean of the response. So let's dig into that Normality assumption a bit. A Normal distribution is symmetric, implying that the mean and the median are the same. As you know, regression estimates changes in the mean of the response.

If you have log-transformed the response, the mean on that scale back-transforms to the median on the original scale. Here is a cartoon to illustrate. Imagine three values from data that have been log-transformed: 1, 2, 3. The mean and median both equal 2. Supposing base 10, these back-transform to 10, 100, 1000. The median is 100, but the mean is 370. You can see that the mean on the log-scale back transforms to the median, not the mean, in the original scale.

OK. Down to business. For the sake of comparison, we will start with details for the case when neither variable has been transformed. And, for simplicity, we will use 10 as the base for the logarithms. If you like to use e, replace "10" in the exponentiations below by $e \approx 2.71828$.

9.6.1 Neither Transformed

The fitted model for the mean of Y is $Y = b_0 + b_1 X$. A 95% CI for the slope is calculated as $b_1 \pm t_{df,.95} \times SE(b_1)$, where $t_{df,.95}$ captures the middle 95% of the t distribution with df degrees of freedom, which here equals the sample size minus the number of coefficients (including the intercept) in the model. For simple regression, it is $n-2$.

For all except quite small sample sizes, $t_{df,.95}$ will be approximately 2. We will use "2" in this discussion to keep the focus on the essentials of interpretation. Using the geyser data for an example, the slope is 10.74 ($SE = 0.626$), and an approximate 95% CI is $10.74 \pm 2 \times 0.626 = (9.5, 12.0)$. The interpretation of the slope comes from subtraction: $(b_0 + b_1(X + 1)) - (b_0 + b_1 X) = b_1$. Here, we are 95% sure the average waiting time increases by between 9.5 and 12 minutes for each minute of increase in the duration of the previous eruption.

9.6.2 *Y* has been Transformed

The fitted model is $\log(Y) = b_0 + b_1 X$, which back-transforms to $Y = 10^{b_0} \times 10^{b_1 X}$. Here, to interpret the slope, we must divide, not subtract: $Y = (10^{b_0} \times 10^{b_1(X+1)}) / (10^{b_0} \times 10^{b_1 X}) = 10^{b_1}$. We can make a CI for the slope on the log-scale in the usual way ($b_1 \pm t_{df,.95} \times SE(b_1) = (LB, UB)$), where LB and UB are the respective lower and upper bounds on the CI. If we back-transform those limits, we get a CI for 10^{b_1}: $(10^{LB}, 10^{UB})$.

For the hormone data from section 10.1, the slope equals -0.095 ($SE = 0.015$), yielding an approximate 95% CI of $(-0.125, -0.065)$. Back-transforming, we get $10^{-0.095} = 0.80$, and $(10^{-0.125}, 10^{-0.065}) = (0.75, 0.86)$. The back-transformed CI is not symmetric; the upper bound will be further from the estimate than the lower bound. In the case of a negative relationship, as we have here, it is easier to discuss the effect of the predictor by changing the scale to a percentage and discussing the size of the drop. We estimate that *median* hormone levels drop by 20% for every advancing year of age, and we are 95% sure the true drop is between 14% and 25%.

9.6.3 *X* has been Log-Transformed

The fitted model is $Y = b_0 + b_1 \log(X)$. For every one-unit increase in $\log(X)$, the mean of Y changes by b_1. Given base 10, a one-unit increase in $\log(X)$ is a tenfold increase in X. Pretty simple, with a wrinkle and an opportunity, both of which we will introduce below. For the species richness versus area data of section 10.2, the slope is 12.3 (SE = 1.76), giving us an approximate 95% CI of (8.8, 15.8). For every tenfold increase in area, we are 95% sure the mean number of species increases by between 8.8 and 15.8.

Here's the wrinkle. What if you had used the natural logarithm (base *e*) instead of 10? The slope in that case is 5.372. So, for every 2.71828-fold increase in area… This is simply awkward. We illustrated in Sections 9.2 and 9.3 how to work with that.

9.6.4 Both are Log-Transformed

This might seem like the most complicated setting, but the arithmetic oddly gets a little easier than the previous setting. The model is $\log(Y) = b_0 + b_1 \log(X)$, which back-transforms to $Y = 10^{b_0} \times X^{b_1}$. Here, we need to discuss changes in both *X* and *Y* in relative terms. Let's use *F* again for your choice of fold-increase in *X*. Then $10^{b_0} \times (FX)^{b_1} / 10^{b_0} \times X^{b_1} = F^{b_1}$. This is the size of the multiplicative change in the median of the response.

In the fish morphometrics data (section 10.3), both weight and length were log-transformed, using base 10. Fish lengths in the sample range from 44 to 263. Tenfold changes are unwieldy, and there are only a few doublings. Let's use 50% increases, implemented by *F* = 1.5. In the fitted model, the slope was 3.04 (SE = 0.016), so a 95% CI is (3.01, 3.07). Now then, for a 50% increase in length, the median weight will increase by a factor of $1.5^{3.04} = 4.56$, and we can be 95% the true change in median weight is between $1.5^{3.01} = 4.5$ and $1.5^{3.07} = 4.6$. These intervals are very narrow because, on average, the relationship between fish weight and length is highly predictable, and so the slope has a very small standard error.

9.7 Chapter Summary

We illustrated, with detailed examples, that logarithmic transformations, when needed, frame relationships in terms of relativity rather than additive change. That relativity might be in the response variable or the predictor, or both. The choice of base is arbitrary and does not affect the relationship.

When log-transforming the predictor, it may be more appealing to discuss relative increases in the predictor at scales different than your choice of base. We show how to implement those choices.

Finally, the arithmetic involved in making inferential statements about the effect of increased values of the predictor, as captured by the slope, can be somewhat mysterious. We lay it out in detail for all the relevant scenarios (neither transformed, Y transformed, X transformed, or both).

Notes

1 Other issues might be in play in certain analyses: you might need to include interaction terms (more on which in the chapter on multiple regression) or need a quadratic term (i.e. include X^2 in the model).
2 Species richness is defined simply as the number of species present.
3 Fisheries ecologists routinely use 10 as the base for log-transformations.
4 You also might see this called a "double-log" model.

References

Dauwalter, D.C, M.A. Baker, S.M. Baker, R. Lee, and J.D. Walrath. 2022. Physical habitat complexity partially offsets the negative effect of Brook Trout on Yellowstone Cutthroat Trout in the peripheral Goose Creek subbasin. Western North American Naturalist 82(4): 660–676

Hillary F. Huber, C. Li, D. Xie, K. G. Gerow, T. C. Register, C. A. Shively, L. A. Cox, and P. W. Nathanielsz. 2024. Female baboon adrenal zona fasciculata and zona reticularis regulatory and functional proteins decrease across the life course. GeroScience, 46(3): 3405–3417. https://doi.org/10.1007/s11357-024-01080-9

Newmark, W.D. 1986. Species-area relationship and its determinants for mammals in western North American national parks. Biological Journal of the Linnean Society 28: 83–98.

Part III

Multiple Regression

Many statistical analyses involve multiple predictors, some of them numerical, perhaps some categorical. In this section, we will introduce you to the intricacies of multiple regression, including how to handle the very common fact of correlation among predictors, how to create and use indicator variables for including categorical predictors, and how to handle the phenomenon whereby the effect of one predictor might change depending on the level of another. This is called an interaction between the predictors, and the presence or absence of an interaction is unrelated to whether the two predictors are correlated.

Variable selection is an important aspect of building a multiple regression model. We show you ways to approach that subject and share our perspective. Here, it is particularly important that you keep your primary goals in mind. Is it an explanation? You should prefer simpler models, perhaps at the expense of prediction. Or is it indeed a prediction, in which case, more complicated models do no harm.

While multiple regression modeling is more complicated than simple linear regression (SLR), you can rest easy about the model assumptions. They are the same as for SLR and can be tested just as easily, using the tools we have already introduced to you. Relativity in the relationships can be a feature of these models just as it can be for SLR, but handling it is no more complicated than we laid out in Chapter 9.

We will introduce multiple regression in Chapter 10, including some of the features listed above. In Chapter 11, we work through several illustrative examples, intended to showcase the flexibility of these models. Chapter 12 is devoted to two essays, one of which elucidates a useful metaphor for understanding multiple regression. There are instructive points of analogy between people working together on a project and multiple predictors

DOI: 10.1201/9781003609605-12

enlisted in a regression model. The second essay is essentially a soap-box essay, our plea for not making models more complicated (or complicated-sounding) than is needed. Finally, we introduce logistic regression, a regression method that is designed for a categorical response variable. We focus on the simplest case, where there are two categories (success/failure, present/absent, and the like).

10

Introduction to Multiple Linear Regression

In this chapter, we spend time on each of the following important elements.

- Creating indicator variables from categorical predictors
- Model selection criteria
  - R^2 and its relatives
  - Mallow's C_p
  - Information criteria
- Selection methods
  - Best subsets
  - Forward, backward, and stepwise
- Interaction terms
  - How to create and use them
  - A publishable error: using correlation to test for interactions
- Multicollinearity
  - What it is and how it is measured; this is relevant when explanation is your goal
  - When it doesn't matter (short version: when prediction is your goal)

This is a longish list of dimensions to address when doing multiple regression. Fortunately, checking the assumptions is identical to what we showed you for SLR (Chapters 6 and 7), and consideration of logarithmic transformations to capture relativity in relationships has largely been covered (Chapter 9).

We will work through an example to illustrate the issues but then follow up on them in more detail in Chapter 11.

10.1 The Terms of Engagement

Multiple linear regression… with more than one predictor, several new concepts come into play. Some level of understanding of each of them will make your modeling much more interesting and informative. Here is a brief introduction to each of them.

DOI: 10.1201/9781003609605-13

An old friend: checking the assumptions. Your reward, if you learned about assumption checking in Chapters 6 and 7, is that *it is exactly the same here* and can be done with the same tools. None of it has changed one little bit.

Scale of measurement. Judicious choice of measurement units can enhance the story-telling. Also, log transformations can be quite useful in capturing relativity in relationships. We explored that in depth in Chapter 10, and you can expect them to be a usual part of your statistical life now.

Purpose: explanation or prediction. Science research is usually focused on explaining the effect of some predictors on a chosen response variable. Sometimes, though, prediction is of primary importance. This choice should affect how you approach the modeling exercise. If explanation is your goal, you might be inclined to favor simpler models; if it is prediction, then you might not care if the model is complicated and difficult to explain. We will mention here that many AI tools focus on prediction almost exclusively (which is fine) without pointing out the consequences of the choice (which is not so fine).

Simply put, the coefficient of determination (R^2) measures the amount of variation in the response that can be attributed to the model. Every time you add another predictor, two things happen.

1. R^2 will go up. Maybe only a little, maybe a lot. But it will go up.
2. The complexity of the model will go up. Maybe only a little, maybe a lot. But it will go up.

So we can discuss this explanation/prediction dichotomy in terms of R^2. The choice is yours: do you prefer a simpler model, paying for it with a smaller R^2? Or do you prefer better predictability, paying for it by having to deal with a more complicated model?. And, as we will see below, your choice should affect your attitude toward multicollinearity and, to some extent, interactions in the model. But first, you need to learn what those things are.

Categorical predictors. We learned in studying SLR that both the response and the predictor need to be meaningfully numerical. We will learn here how to effectively and properly incorporate categorical predictors into a regression model. It is not overly difficult to do so, but some care and thoughtfulness will be helpful.

Interactions between predictors. It is best to start with a conceptual example. In the mountains of the Western United States, small rodents spend much of the winter napping in their burrows. They will rouse themselves from time to time to search for seeds left lying on the ground from the previous summer. They tunnel through the base of the snowpack. Of interest here is the temperature in those tunnels, and the effect of snowpack and air temperature on that.

Imagine the snowpack is only one foot deep. In that case, the effect[1] of air temperature will be dramatic. As the air temperature goes down, so will the temperature in the tunnel. Instead, now imagine there is ten feet of snow insulating the tunnel. The air temperature will have almost no effect; the slope will be virtually zero. Conclusion: The depth of snowpack influences the effect of air temperature.

Now imagine that the air temperature is −20°F. What is the effect of snow-pack on the temperature of the tunnel? Up to a certain point, it will be very strong. Six inches of snowpack might predict a low tunnel temperature, while two feet would predict a warmer. On the other hand, suppose the air temperature is 40°F. Snowpack depth will have little effect. A little insulation? A lot? Tunnel temperatures won't change much. Conclusion: The level of temperature influences the effect of snowpack depth.

When the numerical level of one predictor influences the effect of another, and vice versa, we say there is an interaction between the two predictors. Notice what this does *not* say. It does not say they are related to each other; i.e. it does not say they are correlated with each other. We will come back to that.

Multicollinearity. It is often the case, especially with biological data, that the predictors are correlated with each other. There are many possible patterns of such correlations, but to start, we need to learn when to care about it and how to measure it when we do care about it.

Variable selection[2]. Suppose you have ten different predictors on hand for a given model. How many of them, and which ones, should you include in your model? We will study ways to approach that.

There is no single best path through these topics, and landing on a model can be a dance. Textbook presentations and journal articles are usually very tidy and logical in their presentation. Deductive logic: given this, then that happens. Given that, then this happens. Ken distinctly recalls a moment from his graduate student days when his department hosted biologists to present issues they were wrestling with.[3] He was stunned to see his respected professors flail and wail as they tried to sort out the problem. That was the moment he realized that problem solving is a messy process. Dead ends... Three steps forward, then one or two back. Or sideways.

Every step will be accompanied by the spirit of Tyler Johnson (remember him?). Every choice is a matter of judgment and, therefore subject to criticism. Critical thinking on your part and attention to your purpose will see you through. If the process you will read about now is not neat and tidy, we hope that at least it reflects the sort of process you might need to engage with to find the right model.

10.2 An Illustrative Example: Introducing the Use of Categorical Predictors

For three different crab species, Yamada and Boulding (1997) studied the pinching force (measured in Newtons) of their claws using the height (in millimeters) of their claws as a predictor[4]. Here, for simplicity, we will use just two species; we can call them A and B (model output is in Table 10.1). This categorical predictor can be incorporated into a regression model by using 0 and 1 to code it. This gets the reasonable moniker, "indicator variable." We will use a single column of values, setting 0 to indicate species A and 1 to indicate species B. Some points to make here:

1. The category designated with a "0" is called the baseline category. As you recall, the interpretation of a slope coefficient is along the lines of, "For every one-unit increase in X, there is an estimated (insert slope here) change in the mean of Y." For an indicator variable, there is only a single one-unit increase, from 0 to 1, so that element of the model speaks to the difference in average values of Y between the two categories.

2. You are free to choose either category as the baseline. This choice won't affect the statistical properties of the model, but it might affect the ease of telling the story. There might be a natural story-telling choice. If one group is, say, a control group and the other a treatment group, it might appeal to use the control group as the baseline. A negative slope coefficient might be awkward to discuss; in which case, simply choosing the other category as the baseline category will flip the sign of the coefficient.

3. We won't name the column with the indicator as "species" because by this time tomorrow, we will have forgotten which was coded as 0 and which as 1. Our habit is to name the column to remind us which is the "1." Here, we will call it "species B."

4. Most statistics packages will let you declare a variable to be categorical, and then the package will create the indicators for you. That is fine, but you should feel free to take control of that creation. Some packages will use alphabetic ordering and make the first alphabetically to be the baseline. Or they might use the last.

5. The use of indicator variables when you have three or more categories is necessarily more complicated; we will discuss that in section 10.2.1.

TABLE 10.1

Summary of Regression Terms for the Crab Data; the Equation is $F = -4.55 + 1.12 \times H + 11.01 \times B$

Term	Coefficient	SE of Coef	t-Value	p-Value
Intercept	−4.55	3.56	−1.28	0.214
Height	1.12	0.43	2.62	0.016
Species B	11.01	1.97	5.58	0.000

When we hand these data to students, the first thing they, of course, want to get to the punchline. In that spirit, here is an abbreviated summary of the output. Where does this discussion go?

1. Pinching force cannot be negative, but the intercept in this model being negative is not a problem because a height of 0 implies no crab. The intercept is merely a placeholder to set the location for the model.

2. On average, for every millimeter increase in height, pinching force increases by 1.12 Newtons.

3. The variable called "species B" indicates both species A, by being 0, *and* B, when set to 1. In this model, on average, the difference in pinching force between species A and B is 11 Newtons. Species B is the stronger one.

4. Because of the special nature of the indicator variable, we can take things a step further.

   a. Set it to 0. The term associated with species B drops out. Then the model simplifies to $Y = b_0 + b_H H = -4.55 + 1.12H$.

   b. Set it to 1. Now we get $Y = b_0 + b_H H + b_B = -4.55 + 1.12H + 11.01 = 6.6 + 1.12H$.

So, in this model, the slope associated with size is 1.12, but the intercept is different for the two species. At this point, someone raises the fact that we did not check the assumptions. Gosh, we better do that, since the model might be totally bogus, depending on what we see there (Figure 10.1).

Oh, dear. The goodness of fit is terrible: the curvature in this graph screams at us. At this point, we no longer care about homoscedasticity nor whether the residuals are Normal enough, since the model is clearly invalid. What to do?

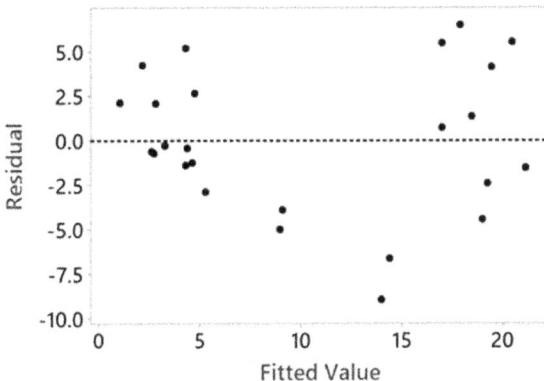

FIGURE 10.1
Residuals versus Fits for the crab strength data.

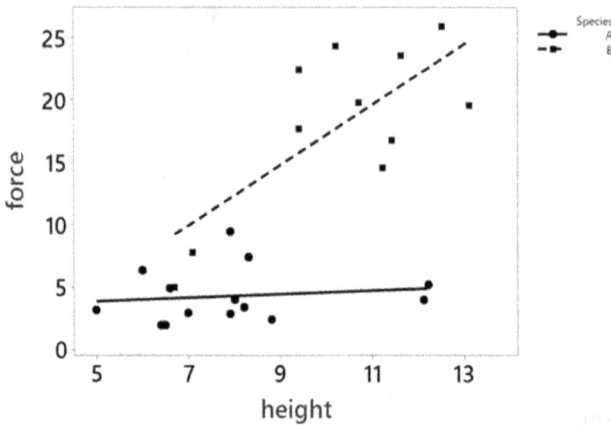

FIGURE 10.2
Scatterplot of crab strength data showing separate regression lines for the two species.

One reason we are using this problem to kickstart multiple regression is that we can get useful information from a scatterplot of the data, so long as we use different symbols for the two species. The result is shown in Figure 10.2. Foreshadowing where this discussion will go, separate regression lines are fitted to the two species.

We see here that the effect of height is different for the two species. Further, the effect of species changes depending on height. For small crabs, the pinching force is about the same for the two species. For large crabs, species B is clearly the stronger one. The solution we will use here is to include an interaction between height and species. To create an interaction term, you need to add a new variable to the data set, created by multiplying together the columns of the two variables of interest.[5] The new column of values usually looks somewhere between meaningless and downright puzzling, but don't worry about that.[6] Let's run the model with that interaction term. This time, before we try to make sense of the numerical output, we will examine the residuals first (Figure 10.3). Learning, we are!

Here is an instructive "unequal scatter" moment. In the scatterplot of the data, species A pinching forces are all quite low. The cluster of values to the left of the graph is mostly from species A. The size of the vertical scatter is about the same on the left side of the plot as on the right. More observations with predicted values around 10 Newtons would have been useful; as is often the case, the data are what the data are, you must live with imperfection (remember Tyler?). The residuals are reasonably symmetric. Let's look at the numerical output (Table 10.2).

It is difficult to make sense of this. The interaction term is significant for any alpha greater than 0.005, meaning that the effect of height is different for the two species, but nothing else is! Or so it seems.

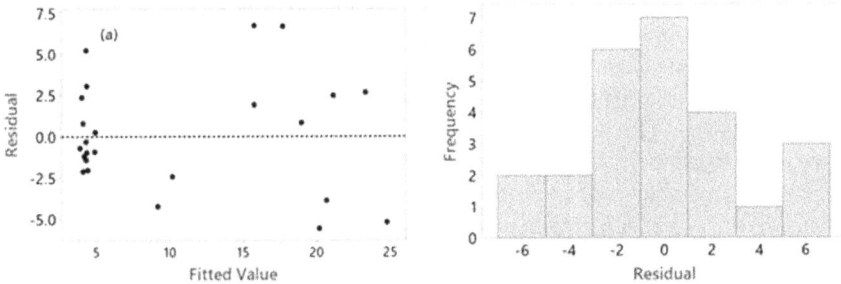

FIGURE 10.3
Residual plots for the crab model that includes species, height, and their interaction.

Before we examine this, a couple of comments are in order. First, we usually don't care about the numerical value of the intercept. Second, it is usual that in a multiple regression, you should feel free to remove from the model any predictors that are not significant and rerun the model without them, which is reasonable if we had three unrelated predictors. Here is a rule, one of very few: if you have an interaction in the model, you must keep the originator terms, no matter if they appear to be significant or not. In this case, their apparent insignificance is misleading. The coefficient for each is the coefficient *when the interacting variable is set to zero*. On the other hand, if an interaction term is *not* significant, also ignore what the computer is telling you about the parent terms. Simply put, run the model again without the interaction term.

Cautionary point: As we said, when you multiply two variables together to create an interaction term, you must keep the originating variables in the model. Their *p*-values are misleading if you take them as representing an overall effect. But sometimes you might multiply two variables together in order to create a new entity. For example, the body mass index for a person is calculated by multiplying weight by one over height squared. It is its own thing, and its creation does not necessitate having height or weight in the model. In the cherry tree example (Chapter 11), the product of diameter squared times height (essentially defining a cylinder) can be used to predict

TABLE 10.2

Summary of Coefficients from the Crab Model that Includes an Interaction Term

Term	Coefficient	SE of Coef	*t*-Value	*p*-Value
Intercept	3.16	3.88	0.82	0.424
Height	0.145	0.475	0.31	0.763
Species B	−10.25	6.96	−1.47	0.156
H × B	2.29	0.73	3.15	0.005

the volume of harvestable wood in the tree. It can stand alone and does not require the two originators to be in the model (although, of course, you can entertain them as possible predictors if you wish).

What is the effect of height? Height can be isolated from this equation: $(0.145 + 2.29B)H$. There is no one value for the slope associated with height. If we set B to zero (indicating species A), then we do get 0.145. Aha! The row for height in Table 10.2 is in reference to the relationship between pinching force and height *for species A*, not for height in general. And as we saw in Figure 10.2, there is not much relationship between height and pinching force for species A.

Playing the same game for the effect of species, we see from the model equation the following for "B": $(-10.25 + 2.29H)B$. The effect of species changes depending on height. If we set the height to zero, we get –10.25. This, in fact, is the difference in the intercepts of the regression lines for species A and B.

We can study this in more detail by setting the species indicator to 0 and then 1.

1. Set it to 0. We get $F = 3.16 + 0.145 \times H$.
2. Set it to 1. We get $F = (3.16 - 10.25) + (0.145 + 2.289)H = -7.1 + 2.43H$.
3. Notice that these two equations match the two lines on Figure 10.2. The equation for species A suggests a flat line, with values near 0. The second shows a much steeper slope.

This version does not tell you immediately whether the slope for species B is significant. To get that most easily, rerun the analysis, but this time code species A as 1, and B as 0. Don't forget to calculate the new interaction term. Table 10.3 shows the results.

Careful examination of Tables 10.2 and 10.3 will reveal that they are showing you the same model, but from a different perspective. The effect of height on pinching force is clearly significant for species B: the p-value is 0.000. The residual plots will be identical to those above, since it really is the same model.

It is easy here to see how the effect of claw size differs between the two species. We can dissect the equation from a different perspective to examine the difference in pinching force between the two species. Which species is

TABLE 10.3

Coefficients for the Crab Model Having Set Species B as the Baseline Category

Term	Coefficient	SE of Coef	t-Value	p-Value
Intercept	–7.09	5.78	–1.23	0.234
Height	2.43	0.55	4.41	0.000
Species A	10.25	6.96	1.47	0.156
H × B	–2.29	0.73	–3.15	0.005

stronger? Note that the range of sizes is from $H = 5$ to about $H = 12$. So, let's use 5 and 12 to represent very small and very large crabs, respectively, and see what the effect of species looks like in each case.

For $H = 5$, we get $F = (3.16 + 5 \times 0.145) + (5 \times 0.289 - 10.25) \times B = 3.89 + 1.2 \times B$.

For $H = 12$, we get $F = (3.16 + 12 \times 0.145) + (12 \times 0.289 - 10.25) \times B = 4.99 + 17.2 \times B$.

For small crabs, there is not much difference in pinching force (1.2 Newtons), while for large crabs it is 17.2 Newtons. This illustrates that the effect of species changes with size. Here again is the definition of interaction. Two predictors (X_1 and X_2, say) have an interaction if the effect of one changes depending on the value of the other, and vice versa.

A publishable error. Here we describe a mistake you can make that is sufficiently seductive and subtle that journal article reviewers might miss it. We have seen published articles containing this mistake.

An interaction between two predictors says nothing at all about whether the predictors are correlated. They may be mutually independent; they may covary. Either state has nothing to do with the presence or absence of an interaction. Unfortunately, the term "interaction" is a slippery one: in everyday English, an interaction between two entities implies some kind of direct relationship. A relationship between two predictors (switching back from English to multiple regression lingo) implies they are correlated. You might be lured into thinking that an interaction between two predictors implies we are discussing a possible correlation between them. Not so. Emphatically not so! The mistake? "We tested for statistical interactions between predictors by estimating the pair-wise correlations among all pairs of predictors. We removed from consideration any predictor that was more correlated than XXX." And here they specify some cut-off of correlation. What they *are* looking at (and even then, getting it wrong) is an aspect of multicollinearity, which is discussed below.

10.2.1 Indicator Variable Coding for Three or More Categories

Let's briefly revisit the situation for two mutually exclusive[7] categories (A and B). Suppose you have two columns of zeros and ones as in Table 10.4.

Every row in your data will either be (1, 0), indicating Category A, or (0, 1), indicating Category B. You can see immediately that using two columns is redundant in that you can discern complete information about category membership from either of the two columns. Which one you use in your analysis

TABLE 10.4

Indicator Variable Coding for Two Categories

Category A	Category B
1	0
0	1

TABLE 10.5

Indicator Variable Coding for Three Categories

Category A	Category B	Category C
1	0	0
0	1	0
0	0	1

is arbitrary but has story-telling implications. In particular, the zero represents what we will call the baseline category. Table 10.5 shows the analogous situation for three categories.

Every row in your dataset will look like one of these three. One column (any, your pick) is redundant. Cover up one: complete information about category membership is to be had with only two of the columns. Here, the row with two zeros is associated with the "left-out" column; it is present implicitly as the "baseline" or "reference" category. For the sake of discussion, let's suppose we decide that the third column will be our reference category (and we will name the remaining two simply as "A" and "B." In the resulting model, the slope coefficient associated with A tells you about the shift of mean values in the response from category C to category A; that for B shows the analogous shift from C to B. Here are a few things to note:

1. However, many categories you have, you need one less than that as indicators.
2. The "left out" category is present in the model implicitly as the so-called baseline category.
3. The slope coefficient for each of the indicated categories shows the shift in mean response from the baseline category to the indicated one.
4. The choice of baseline category is arbitrary; you should choose it to simplify the storytelling as much as possible[8].
5. If you decide to include an interaction between a categorical variable and some other variable, you must form a separate interaction term for each indicated category. You need to keep all of them or none of them in the model; otherwise, the story will become incoherent in one way or another.

10.3 Multicollinearity: Correlation among the Predictors

It is usual in many data sets that predictors are correlated with one another. In field studies, looking at, say, bird abundance as it relates to distance from industrial activity (e.g. oil or gas wells, wind turbines), factors such as density of plants, distance from water, and so on might also affect bird abundance; these things are most likely correlated with one another. Plots that are near water, for instance, might have higher shrub density and, because of being in a low-lying place, be not close to a wind turbine. Another example: in morphological data, physical measurements (height, stomach circumference, bicep circumference, etc.) are usually positively correlated with one another. And so on.

Designed experiments sometimes have uncorrelated predictors. For instance, a greenhouse study might have all combinations of four levels of temperature and three watering levels. These are considered independent of each other in that if you know the watering level, you get no insight into the temperature, since all four will occur with that watering level.

Let's formalize the notion of multicollinearity and examine its consequences. Suppose you ran a regression using a predictor X_j as the response variable, in a model with all the other predictors as well, predictors. Let R_j^2 denote the R^2 obtained from regressing that variable on all the other predictors. A high value would be evidence of multicollinearity. It might be that X_j is particularly correlated with one other predictor or is diffusely correlated with each of the others, but *collectively* they can predict X_j readily.

Why does this matter (when it matters, see below)? The variance of the slope estimate for the generic predictor "j" in a multiple regression is $var(B_j) = \frac{S_{res}^2}{(n-1)S_j^2(1-R_j^2)}$, where

1. S_{res}^2 is the variance among the residuals,
2. S_j^2 is the variance among the X_j values, and
3. R_j^2 is the coefficient of determination from regressing X_j against all the other predictors.

In this formulation, $\frac{1}{(1-R_j^2)}$ is called the variance inflation factor (VIF). To see why, consider that if X_j is perfectly uncorrelated with the other predictors, R_j^2 would be zero, and the VIF would equal one. But if X_j is highly correlated with the other predictors, R_j^2 would be close to one, so the denominator of the VIF would get close to zero, and the VIF would be very high. Do the math. Set R_j^2 to be 0.99. The VIF would be 100.

10.3.1 When Does It Matter? An Analogy...

Collinearity among predictors is a signal of overlapping information. Suppose X_1 is a good predictor of Y, and that it is correlated with another predictor X_2. It makes sense that X_2 is also a decent predictor of Y, but if you have X_1 in the model, adding X_2 might not gain you much, since its information content is roughly already present in X_1. And vice versa.

Collinearity among predictors analogizes to overlapping skill sets among groups of people assigned to work on a project. If the skill set of one person is nearly identical to that of a second, perhaps you don't need both on the team. On the other hand, it is almost guaranteed that skill sets among team members will overlap to some extent. If your goal in the group project is simply to get a great project report, the overlapping skills causes no problems at all; it could even be an advantage as folks could check each other's work. On the other hand, if you want to understand what each person contributed, then the overlapping skills might get in the way of your assessing those contributions[9].

For example, salinity is a measure of the amount of dissolved ions in water; conductivity measures how those ions conduct electricity. They are highly correlated. One of our students used both as predictors and was surprised at the large standard errors of the coefficients.

Regression models are usually built for one of two reasons: explanation of the role of one (or some) variable(s) in explaining changes in the mean of the response, or, more simply, prediction of the mean of the response.

If your goal is prediction, multicollinearity, if present, can largely be ignored. You don't necessarily care if the model is opaque and difficult to parse. Its predictive performance is more important. Prediction as a goal is not usual in research work, but *if* it is the goal, you can mark yourself as safe from multicollinearity issues.

If your goal is explanation, then multicollinearity might be a concern. The very idea that the SE of a predictor gets inflated by multicollinearity implies that

1. confidence intervals for the slope of that predictor will be wider, and

2. p-values will be larger, making it harder to detect effects.

In most research studies, explanation is indeed of interest, and so learning about multicollinearity is useful. Let's focus on the explanation scenario and consider when, how, and how not to measure multicollinearity. Suppose you have a dataset with ten predictors.

Bad idea number one. We've seen published papers where folks claim to have accounted for multicollinearity by examining all the pair-wise correlations among the ten predictors and removing from consideration those with high pair-wise correlations. Perhaps they will remove one of the two. There are two problems with this. First, looking for multicollinearity in a pairwise manner among all ten predictors is the wrong place to look. Even if all the pair-wise correlations are on the low side, that does not imply that multicollinearity is absent. Recall that multicollinearity for X_j is measured by R_j^2, the coefficient of determination between that predictor and all the others. It is certainly possible that the other nine predictors could collectively predict Xj even though any one of them, taken by itself, does not. **Use the VIF.**

Bad idea number two. The process of building a regression model often includes so-called model selection, which is a process by which certain predictors are retained in the final model and others eliminated. There are different ways to approach this, but in most procedures, the selected variables tend to be variables with less multicollinearity. For example, if X_1 and X_4 are in the model, and X_2 is correlated with them, perhaps X_2 needn't be included. So, bad idea number two is to assess multicollinearity among all ten predictors. That will almost surely not be your chosen model, and you ought to care only about multicollinearity in your chosen model. So, **use the VIF to assess multicollinearity only on models under full consideration.**

Multicollinearity and interactions. If you add an interaction term to the model, there will be a resulting increase in multicollinearity, by necessity. Considering that the interaction term is created by multiplying together the values of the two "parent" terms, then it will tend to be positively correlated with both. In this case, multicollinearity must simply be lived with...

10.4 Model Selection (Better Called "Variable Selection")

Model selection is the process of selecting which and how many predictors to include in a model. Often, studies are done with a particular focus. "Is bird abundance affected by presence of wind turbines?" "Might this treatment improve condition X?" If you are in a situation where you have measured many variables that might be related to the response variable, you then have the problem of choosing among them. Should you favor models with fewer predictors (less precise predictions can be made, but the model is simpler to understand) or models with more predictors (better predictions, but messier models)? Once you have decided on your inclination in that regard, how do you decide where to stop adding predictors to, or removing predictors from, your model?

Model selection should not usually be done on statistical criteria only. For example, a predictor may be in a model because of theoretical considerations. For another, you might wish to compare the model to others in the literature. A predictor that is very expensive to measure might be included in a model only if its utility outweighs the expense. And etcetera. All we can deal with here are the statistical elements; keep in mind that they are tools, not rules.

There are two dimensions to the model selection process that we will now discuss.

10.4.1 Summary Criteria by Which to Judge a Model

Some of these are of the "less is more" school of thought (values built on residual variation: smaller is better); two are of the "bigger is better" school (values built on the predictive ability of the model). Most have a built-in penalty for having too many predictors; some penalize more than others.

10.4.2 The Coefficient of Determination: R^2

The simplest and most often used criterion for the predictive quality of a model is R^2. This is the proportion of the variation of the response variable that is explained by the model; it was introduced in Chapter 6. This criterion is typically reported as part of the summary of a regression model, in part because of its simple interpretation. If you were to add a new predictor to an existing model, the R^2 value will go up. It may go up only a little, if the new predictor adds almost nothing to the predictability of the model, or it may go up a lot, if the new predictor is indeed useful, and the model is not already predicting very well.

10.4.3 Adjusted R^2

A problem called "over-fitting" occurs with data sets having a small number of observations relative to the number of predictors. Adjusted R^2 tries to account for this phenomenon by reducing R^2 according to the number of predictors in the model. A large drop from R^2 to adjusted R^2 indicates possible over-fitting (i.e. too many predictors given the sample size).

10.4.4 Mallow's C_p

On the assumption that the model with all the available predictors accounts for all possible predictions, the MSE[10] from that full model represents truly random error. A model with fewer predictors will leave more unexplained variation; this criterion will yield a larger value. Among models with p predictors, a smaller C_p is considered to point to a better model.

10.4.5 Information Criteria

Two other criteria to consider are Akaike's Information Criterion (*AICp*) and Schwartz' Bayesian Information Criterion (*SBCp* or BIC_p). The subscript references the number of predictors, but (for the sake of simplicity here), not *which* predictors. AIC is computed as $AIC_p = n\ln(SSE_p) - n\ln(n) + 2p$. For a given full data set,

1. larger SSE implies more "random error," and increases AIC, and
2. more predictors increase AIC by increasing the third term in the formula.

AIC is available routinely from many statistics packages, or you can get the SSE from the ANOVA output; sample size (n) and a few minutes with your calculator will get you there.

The second one is computed as $SBC_p = n\ln(SSE_p) - n\ln(n) + \ln(n)p$. You can see that the formula is quite like that for AIC. Comparing two models, the one with the smallest AIC/BIC would be considered the better of the two. AIC values can be negative, so a value of -10 registers as smaller than, say, -1. BIC puts greater emphasis on parsimony. A cautionary point: AIC used to compare a model with log-transformations for the response to one without such will be misleading because of the rescaling

What do we use? We use R^2, with a glance at adjusted R^2, keeping in mind the purposes for doing the modeling. Do we have a reason to prefer fewer predictors, or do we favor predictability itself, at the possible expense of simplicity of the model? We do not favor using any one criterion in a rule-bound manner. *You* must take charge of your modeling...

10.4.6 Model Selection Strategies

There are several selection strategies that are now of mostly historical interest (forward selection, backward elimination, stepwise selection, and variations thereof); we won't discuss these here. A method currently in popular use is called "best subsets selection". It is the only one that is guaranteed to consider all possible combinations of the variables. Most statistics packages will give a summary of the best two models comprised of singletons, the best two models with only pairs of predictors, the best two triplets, and so on. The output for each model usually consists of R^2, R^2-adj, and possibly another metric or two. You need to stay in charge of the process: best subsets is driven by arithmetic alone, and it may come up with models you *don't* prefer due to their biological implausibility. Use it to generate lists of possible predictor teams for further analysis.

10.5 Chapter Summary

Of the main topics we need to consider in multiple regression, we began with categorical predictors and how to work with interactions. It is conceptually straightforward to incorporate categorical predictors, especially if, as in the example in this chapter, there are only two categories. This is done by using so-called indicator variables, wherein 0 and 1 indicate the two categories.

Interactions between predictors introduce complexity in a model, which you will either find quite interesting or unhappily complicating. You can choose to ignore them, but as the example we used here illustrates, doing so might sometimes miss very important features of the data. We pushed our way through all the details for the crab strength model to give you an idea of how to proceed with your own data.

An interaction term is created by multiplying together the values of the two variables in question. A rule arises when incorporating an interaction: if you keep the interaction in the model, you should keep the original variables as well, whether they appear to be significant or not. They are necessary to allow the complications associated with the interaction to manifest reasonably.

Correlation among the predictors, a.k.a. multicollinearity, is an omnipresent feature of data sets (except, occasionally, in designed experiments. It can be ignored if prediction is the main goal; otherwise explanation) use the VIF on models of interest. Variable selection (discussed here also) reduces the number of predictors and usually does so in a way that reduces multicollinearity.

Notes

1 Throughout our discussion of multiple regression, when we mention the "effect" of a predictor, we are referencing the slope associated with that predictor.
2 Some people call this "model selection." I don't favor that term since it implies that once you have done this stage, you are done. To be sure, choosing which variables to include is a critical part of model building.
3 Jorge has had similar experiences
4 When crabs walk along the ocean floor, hunting, they raise their claws up into the water above them. Yamada and Boulding referred to the "height" of the crab claws; we will use that term here.
5 These days, many statistics packages will do this work for you, if you call for an interaction.
6 In this case there will be a bunch of 0 values (coinciding with Species B = 0) and another bunch of values of height coinciding with Species B = 1.
7 Mutually exclusive means that if you are in any one category, you can be in no other. Without that, this indicator business falls apart.
8 Sometimes it will pay to repeat the analysis with different baseline choices to see which choice bet enhances the tale.
9 Of course, you could ask the team members to report on what they and others contributed. Since regression predictors can't do that, we will leave that idea out of the discussion ☺.
10 The mean squared error in the ANOVA is the estimated variance among the residuals.

Reference

Yamada, S. B., and E.G. Boulding. 1997. Claw morphology, prey size selection and foraging efficiency in generalist and specialist shell-breaking crabs. Journal of Experimental Marine Biology and Ecology 220 (1998): 191–211.

11

Multiple Regression Examples

Given our belief that practice with the tools is what enables deeper learning, we will devote this chapter to working through some illustrative examples of multiple regression.

The first example is a botanical mystery. At first, it seems a simple model works (assumptions all fit), but things fall apart upon closer examination, with an unexpected conclusion. Model creation seems uniquely to be a multiple regression issue. We return one more time to the geyser data to show that deeply honoring the purpose of modeling can lead to a counterintuitive choice (teaser: it includes tossing out all the statistical models!). And finally, Tyler Johnson on steroids: how many statistically acceptable models can one make with only two predictors? We answer that question (there are more than you might think) using an example of predicting cherry tree wood volume given the diameter and height of the trees.

We suppose that by now you understand that doing a statistical analysis is not a step-by-step procedure (Step One, Step Two, Step Three, Done). It is often a dance, a few steps in this direction, then that. It is an exploration of possible ways to shed light on the underlying story. As we have learned the hard way, simply explaining ideas to students and then testing them on the ideas does not leave many of them adequately equipped to forge ahead on their own. Deep learning of the application of statistical methods requires repeated practice. We cannot give you a huge amount of that in this text, but in this chapter, we present several examples, each of which demonstrates the process of discovery and highlights Tyler's aphorism. We hope these tales will help you accept and embrace that process with your own data sets and questions.

11.1 A Riddle inside a Mystery inside an Enigma

Winston Churchill used this phrase to describe the intentions and interests of Russia in 1939. It suggests that there is something hidden, impossible to foretell. This example feels like that and really drives home the need to explore your data, graphically if possible.

Dr. Ramesh Vaidya, a botany professor at a rural college associated with Shivaji University, Kolhapur, India, studied the effect of compounds that leach from plants (leaves, bark, fruit) on the germination rates of seeds (data came to us compliments of Dr. Anil Gore). Leachates were collected from the leaves of a species of *Ageritina adenophora*[1] and leaves, bark, and fruit of *Catunaregam spinosa*[2]. Leachate concentrations of 0%, 0.5%, 1%, 1.5%, 2%, and 2.5% were prepared and used to water the seeds of mung beans and sorghum. Does the concentration of the supposed toxin affect germination rates? As a side question, does the source of the leachate matter? And, finally, do the two target species respond similarly?

One might expect germination rates to decline with increasing concentrations, since the leachates tend to be toxic. To the surprise of the researchers, the correlation, although slightly negative, was not significant. The *p*-value for the test was 0.367. It is time to dig deeper.

What follows is from our exploration of the data; we're sure the researchers encountered much of this, but we don't want to speak for them. We ran a model with source (the four leachate sources), seed type of crop (mung and sorghum), and leachate concentrations. The first two are categorical variables, which we will incorporate using indicator variables. Our first model included all three predictors. The residual plots look reasonable (Figure 11.1). There is no strong evidence of lack of fit, scatter looks plausibly equal, and the residuals are not too badly non-Normal, given the sample size of 42.

In fact, this residuals versus fits plot is one of the nicest ones we have seen. A lovely "random cloud" of points exhibiting neither curvature nor unequal scatter. What could go wrong?

It appears that the source of leachate is not important (*p*-value for that factor is 0.602). We suppose that was happily noted by the researchers, as it suggests that any leachate effect in this study is not dependent on the source. There is so far no effect of concentration of leachate (*p*-value for that is 0.332; odd, that), but the two seed types seem to germinate at different rates[3] (*p*-value was 0.017).

Let's take the leachate source out of the model and try again. Whoa! Crikey! Get off the bus! What the heck is going on in that residuals versus fits plot

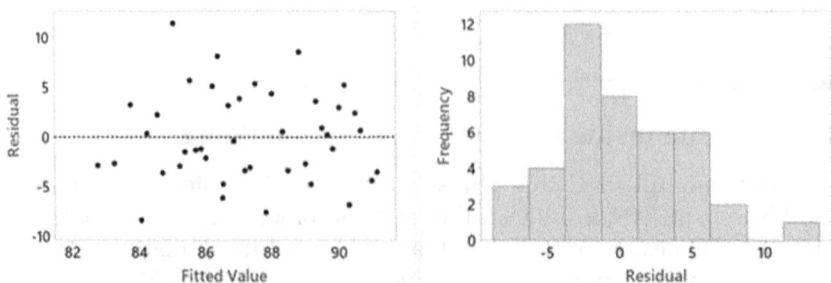

FIGURE 11.1
Residual plots from fitting all predictors to germination rates.

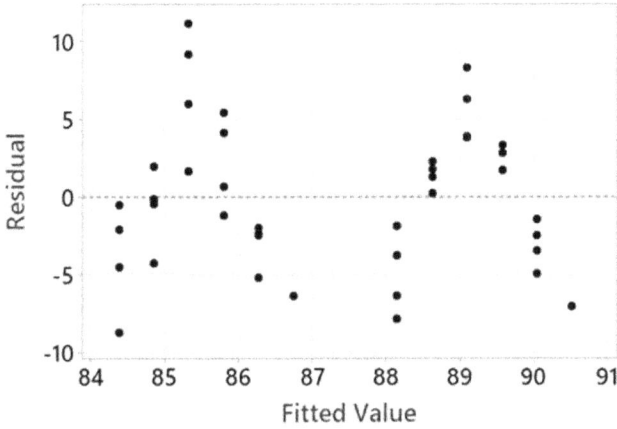

FIGURE 11.2
Residuals versus fits for a model with type of seed (mung or sorghum) and leachate concentrations.

(Figure 11.2)? There is no point in going on to the histogram of the residuals or the numeric output. We have a problem.

Pretty baffling, but the model elements have been simplified now to seed type and leachate concentration. That enables us to look at the raw data in a fairly simple way, since one of the variables (seed type) is categorical with two categories (mung and soybean). Let's take a look (Figure 11.3).

THIS is interesting, and perhaps unexpected. It appears that germination rates go up for very low concentrations of leachate, and then start dropping. As a side note, mung beans seem to germinate more readily than sorghum:

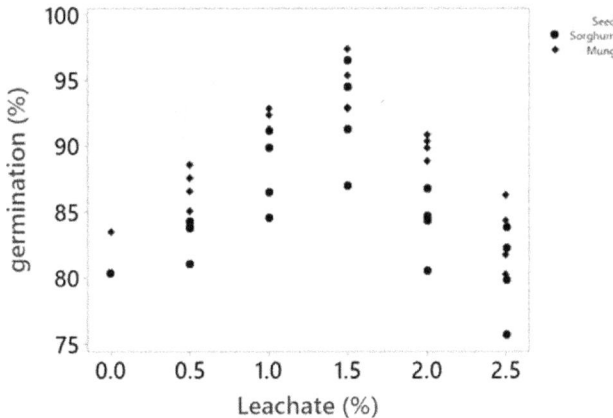

FIGURE 11.3
Germination rate versus leachate concentration, with different symbols for the two seed types.

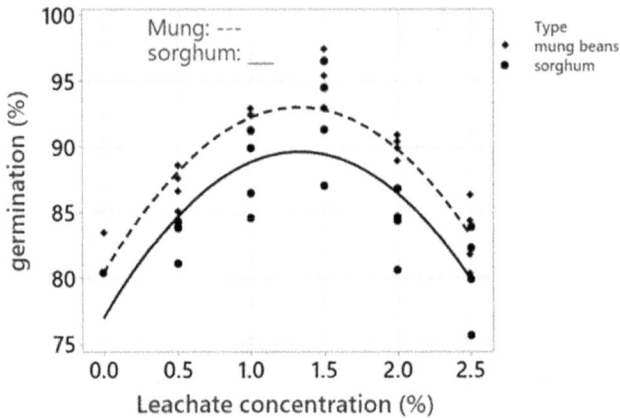

FIGURE 11.4
Fitted curves to the germination data.

at each level of leachate, the mung symbols sit higher on the graph. In our experience, a quadratic model is not often biologically tenable, but these data are an interesting counter-example (Figure 11.4).

This is an elegant and easy-to-discuss model that was practically invisible at the outset. Whatever compounds are in the leachate seem, at very low concentrations, helpful for germination. Fairly quickly, however, they become inhibitory. What was interesting to us was how the residual plots guided us in the search for a model.

Recall that the definition of an interaction between two predictors is that the effect (i.e. slope) associated with one changes depending on the level of the other, and vice versa. Recall also that you include an interaction in a model by multiplying together the observed values of the two predictors in question. A quadratic term for concentration has the form C^2, i.e. we multiplied each value of C by itself. You could think of a quadratic term as an interaction between a predictor and itself. An odd idea, perhaps, but the effect of C does indeed change with levels of C. The instantaneous slope at $C = 0.5$ is quite strongly positive; it is close to zero when $C = 1.3$ and then becomes sharply negative thereafter.

Speaking of choosing models, let's revisit the geyser data one more time...

11.2 Selecting the Best Model

Model selection is, as it ought to be, an important element in multiple regression, but the geyser data present an interesting study in model selection that demonstrates the importance of being true to your intentions. Recall

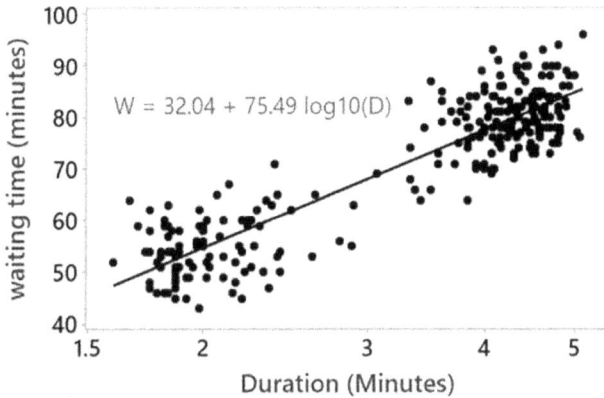

FIGURE 11.5
Fitting W to log(D) for the geyser data.

the problem: as an Upper Geyser Basin park ranger in Yellowstone National Park, you would often be asked, "How long do we have to wait for the next eruption?" The data on waiting times suggest an unfortunately large range. "I am pretty sure it will be between 45 and 95 minutes." A 50-minute range is unsatisfactory. That motivated us, in Chapter 6, to study the relationship between waiting times and durations, leading to the prediction equation $W = 33.5 + 10.7 \times D$. This can be accompanied by plus or minus ten minutes. A 20-minute range is quite better than 50.

Recall that when we were doing a Goodness of Fit test by inspecting the residuals versus fits plot, we mentioned that there was an ignorable hint of curvature (Figure 6.6 if you need a reminder). We can model the presence of that curvature in several ways, but one option is to take the logarithm of the duration values (Figure 11.5).

Note that curvature has been removed, and the distribution of the residuals is quite nicely symmetric. So, in the sense of satisfying the assumptions, it is somewhat better. But now imagine trying to use it.

Old Faithful geyser has just erupted, and you dutifully noted that the eruption was just over four minutes and ten seconds long. Call it 4.2 minutes for the sake of a number. Quick! You can see tourists heading your way who just missed it! What is $\log_{10}(4.2)$?[4] Then multiply it by 75.5 and add 32. Yikes! This won't work. The original simple linear regression model is much simpler to work with; perhaps a small sacrifice of "perfection of assumptions" would be worth it. Quick! What is $4.2 \times 10.7 + 33.5$? This is still a bit daunting to do on the fly, unless you have a calculator in your uniform pocket.

Here is a radical idea: jettison the statistical models and use a simplified version. Quick! What is $4.2 \times 10 + 35$? With a bit of practice, you could do that in your head even if you are being baked like a potpie in the afternoon sun. Then, say, add "plus or minus 10 minutes" to your answer? How well would

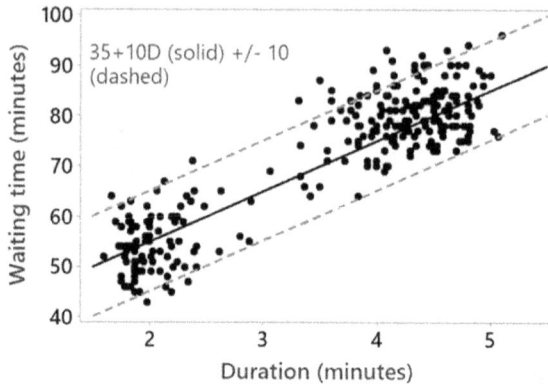

FIGURE 11.6
Rough-and-ready predictions of waiting time.

that work? Figure 11.6 shows the data with this simplified approach on display. Not bad. You could modify this by adding 15 for the upper end instead of 10, but folks won't even notice if they must wait a few minutes longer to see the eruption. They would be MUCH unhappier to come out of the cafeteria, having had lunch, only to find out they just missed it.

We cannot imagine a scientific situation where dismissing a statistical model for something different would be a good idea; in that regard, this situation might be unique. But it does dramatically demonstrate the importance of having a clear reason for doing the analysis and then allowing that purpose to inform your judgment regarding what is best.

11.3 Tyler Unleashed

This is an innocent-looking data set that featured two predictors, height and dbh[5]; the job at hand is to predict the volume of harvestable lumber in cherry trees (Hand *et al*, 1994, page 159). In class, Ken entreated his students to be creative in building their models, and boy howdy, did they ever (Table 11.1)!

Notes:

1. CSA (cross-sectional area) was defined as $CSA = \frac{\pi}{4}D^2$.

2. Some student teams indulged in variable selection. They noted that, given the diameter, height didn't add much. That's why models 3, 4, and 8 include only diameter, in one form or another.

3. Why do models 7 and 9 have the very same R^2? Coincidence?

TABLE 11.1

Summary of Student Models and Their Performance. "Popularity" Notes How Many Teams Settled on that Model. The Assumptions were Adequately Met in Each Case.

Model	Response	Model Elements	R^2	Popularity
1	V	H, D, H·D	97.56	6
2	Log(V)	H, D, H·D	97.69	2
3	Log(V)	Log(D)	95.39	3
4	V	D, D²	96.2	1
5	V	H, D, D², H·D ²	97.79	1
6	Log(V)	H, D	96.84	1
7	Log(V)	H, log(D)	97.76	1
8	Log(V)	D	94	1
9	Log(V)	H, log(CSA)	97.76	1
10	Log(V)	Log(D), log(V)	97.7	2

You understand that if we change measurement units (for some predictor) from feet to inches, all we are doing is multiplying X by 12. We haven't changed the relationship. A graph of Y versus X would look identical, except for the scale of the numbers on the X axis. The model properties would remain the same, except for the numerical value of the slope. What was formerly a "one-unit" change in X (in feet) is now twelve of them. Accordingly, the slope with X in inches would be 1/12th of its former self.

So. $\log(\frac{\pi}{4}D^2)$ (model 9) could be written as $\log(\frac{\pi}{4}) + 2 \times \log(D)$. Model 7, which incorporates $\log(D)$), would yield essentially the same model as model 9 (albeit with different values for the coefficients). In particular, the R^2 values and p-values would be identical.

11.4 Chapter Summary

We hope you have gained the following from these examples. First, the importance of an open mind and creatively exploring the data (Example 1). It is very important to keep in mind your goals for the analysis, as they might lead to modeling choices that might not otherwise occur (Example 2). Finally, the third example illustrates Tyler Johnon's aphorism quite clearly: many different models might do a decent job, and it is up to you to choose the one you prefer.

A rough flow for doing a multiple regression analysis would include the following (in this order).

0. The response variable needs to be meaningfully numerical[6]; predictors can be a mix of some numerical, some categorical.

1. Establish the right scale for the model. You might need to try log-transforming Y or numerical predictors (or both). Sometimes this is a dance step. You might have to try a few models and let the residual analyses give you clues.

2. If the list of variables is long and chosen so as to cover your bases and not miss anything, deploy variable selection to reduce the list to (say) 3 or 4. This can be considered an exploratory phase.

3. Once you have established a few candidate teams of predictors to work with, get creative. Consider interactions, quadratics... keep your original goal (prediction/explanation) in mind.

We emphasize log transformations since they capture relativity, and relativity in relationships is almost ubiquitous. There are other transformations out there, but

1. They often don't work much better than log transformations, and

2. They often don't lead to straight-forward storytelling. If the statistics gets in the way of the biological story, we ask, "What's the point?"

Notes

1 The genus is known as snakeroot, the roots of which were used by Native Americans to treat snake bites.

2 Mountain pomegranate

3 Comparing mung and sorghum germination rates is not a primary focus of this study; the effect of leachates was being studied.

4 It is 0.623, for all the good that will do you.

5 Traditionally, foresters measure tree diameter four feet above ground; dbh stands for "diameter at breast height."

6 Logistic regression is designed for the case of a categorical predictor (e.g. success/failure, presence/absence). We will study that in Chapter 13.

Reference

Hand, D.J., E. Daly, A.D. Lunn, K.J. McConway, and E. Ostrowski. 1994. A Handbook of Small Data Sets. Chapman and Hall. 458 pages.

12

Two Essays on Multiple Regression

In this chapter, we share two essays on multiple regression. In one, we show analogies between elements of having people working together on projects and multiple regression; people and predictor variables playing analogous roles. There are many things that we understand intuitively about people working in groups. For example, adding more people to the team will help up to a certain point, and then not so much thereafter. Overlapping skill sets among partners won't cause the final product to worsen, but it could make it difficult to divine the contribution of each group member. This analogizes to overlapping information content between predictors (i.e. correlation). We hope that this metaphor might help make multiple regression matters more intuitive.

In the second, we go to the heart of a foundational premise of our approach to statistical modeling, as indicated by the aphorism in the last sentence of the preface: "Don't let the statistical tale wag the biological dog." We demonstrate by way of an example that complicated approaches are not always the same as sophisticated approaches and can sometimes obscure the story the data have to tell.

12.1 Multiple Regression without the Math: A Conceptual Analogy

A first introduction to multiple regression can be daunting, particularly for students of the sciences who need to learn how to use it but might well not take courses to develop their matrix algebra skills, which is required for a more mathematical presentation. How do you help students deeply understand the roles of multicollinearity, interactions between predictors, model selection, and so on, without the math? A team of people working on a project stands as a rich and powerful conceptual analogy to a multiplicity of predictors in a regression model. The analogy is rich because almost every major topic in multiple regression has an analogue when considering a team of people working on a project; powerful because we intuitively understand many aspects of teamwork, and that understanding transfers readily to new understanding of analogous aspects of multiple regression.

DOI: 10.1201/9781003609605-15

We will begin by describing a simple multiple regression problem that we will use as an example throughout. Various body measurements were taken on 242 human subjects with the goal of using some combination of them as a predictive model of weight in kilograms. Here, we are going to consider only three of the predictors: bicep (B), forearm (F), and wrist (W) circumferences, each in cm.

In what follows, we will dwell on how the slopes, their standard errors (SE), the model R^2, and other features of the regression model change with choices of predictors to include in the model, which analogizes to the choice of people to place in a team. For each topic, we will develop the relevant analogy to teams of people and show how what we know about teams of people translates directly into understanding the working of multiple regression.

12.1.1 Slopes: Interpretation and Behavior

Simply put, we estimate the mean of Y in a multiple regression with an equation of the form $\hat{Y} = b_0 + b_1 X_1 + \ldots + b_p X_p$, where b_0 is the estimated model intercept and b_1 through b_p are the estimated slope coefficients for the p predictors. If you have three people from which to choose a team for some project, the final product and the contribution of each person will change depending on how many (and who) are in the group. The effect of any one individual will not stay the same when their team changes. In multiple regression, the analogue is the slope associated with a predictor. For instance, the slope coefficient for bicep circumference is different depending on which model it is part of (Table 12.1). Notice also that the numerical value of the slope for biceps circumference tends to be less as the team gets larger. This would analogize to an individual having to work less in a larger group. You know it won't always happen that way, and so it is with multiple regression: the slope coefficients

TABLE 12.1

Model Equations for Possible[1] Models to Predict Weight from Biceps (B), Forearm (F), and Wrist (W) Circumferences (all in cm)

Model	"Team"	Predictive Model
1	Biceps circumference	Weight = −26 + 3.33B
2	Forearm circumference	Weight = −55 + 4.72F
3	Wrist circumference	Weight = −94 + 9.66W
4	Biceps and forearm	Weight = −5 + 2.19B + 2.14F
5	Biceps and wrist	Weight = −85 + 2.36B + 4.82W
6	Forearm and wrist	Weight = −100 + 3.20F + 4.86W
7	All three	Weight = −87 + 1.94B + 1.17F + 4.03W

Note: Primary interest is in the slope coefficients, so we rounded the intercepts to a whole number.

TABLE 12.2

Slopes and Their Standard Errors for the Seven Models of Our Weight-Modeling Example

Model	"Team"	Slope (SE) Bicep	Forearm	Wrist
1	Biceps circumference	3.33 (0.16)		
2	Forearm circumference		4.72 (0.26)	
3	Wrist circumference			9.66 (0.61)
4	Biceps and forearm	2.19 (0.25)	2.14 (0.37)	
5	Biceps and wrist	2.36 (0.18)		4.82 (0.58)
6	Forearm and wrist		3.20 (0.32)	4.86 (0.70)
7	All three	1.95 (0.23)	1.17 (0.37)	4.03 (0.62)

aren't guaranteed to get smaller with increasing size of predictor teams, but they often do.

12.1.2 Precision of the Slope

The SE of a statistic is a measure of the precision associated with that statistic. For almost every case for each predictor in our example, the SEs go up when you move from a model with only one predictor to a model with two predictors to a model with all three (Table 12.2). If you have three people in a group, it might be harder to assess the contribution of any one individual to the project than if the person were working by themselves or with just one other. It tends to get harder to be sure of exactly what contribution is made by any given predictor (i.e. the SEs will be larger) for reasons that are, by analogy, like the reasons it gets harder to assess the role of an individual person in a large group (more on that below).

12.1.3 Overall Product: R^2

Recall that R^2 is defined as the proportion of the variability in the response that is explained by the regression model. If you have three people in a group, you might expect to get a typically better product than you would get from a group of only two of them or one of them working alone. You also understand that there is a "law of diminishing returns": at some point, adding more people will not appreciably improve the product. For this regression example, the case of diminished returns occurs when moving from a set of two predictors to a set with three predictors (Table 12.3). Every multiple regression problem is different, and so it is with teams of people and projects; sometimes four or five predictors are desirable, occasionally, a single predictor is sufficient.

TABLE 12.3

Coefficients of Determination (R^2) for
the Seven Models

Model	"Team"	R^2 (%)
1	Biceps circumference	64.2
2	Forearm circumference	58.4
3	Wrist circumference	50.9
4	Biceps and forearm	68.7
5	Biceps and wrist	72.3
6	Forearm and wrist	65.4
7	All three	73.4

With R^2 as a measure of "quality of model," we see that when the predictors work alone, biceps circumference does the best job, and forearm circumference is second best. When they are put into pairs, it is the combination of biceps circumference and wrist circumference that is best. This may happen with teams of people, also. The two people judged to be the best and second-best to be assigned to do the project by themselves might not be the best pair of people to do the work. In this regression case, the best pair does a notably better job than the best singleton (R^2 goes up from 64.2% to 72.3%. Adding a third member to the team does not make such a difference: the coefficient of determination for the model with all three predictors is 73.4%, only a modest increase over the job done by the best pair (Table 12.3).

12.1.4 Correlated Predictors and Over-Lapping Skill Sets among Team Members

It is usual in most data sets arising from observational studies that the predictors are at least somewhat correlated. Suppose two people with highly overlapping skill sets are put in a group together to work on a project. Is that a good idea or a bad one? There is no single answer. The pair may very well deliver an excellent product (putting them together may be a good idea if the main goal is a great product); on the other hand, it may be hard to tell precisely who contributed what to that product (so it may not be as good an idea if assigning "individual grades" is important).

For each predictor, we can consider the R^2 value when it is regressed against the other two predictors. The largest such is for forearm circumference (69.7%), then biceps (64.2%); the smallest is for wrist circumference (48.4%). The SEs for the slopes associated with wrist circumference change the least when other predictors are added to the model, whereas both bicep and forearm SEs show more dramatic changes. Here, the correlation between bicep and forearm values is reflected in a certain degree of uncertainty regarding the effect of either when both are in the model. Here, we are seeing the effects of multicollinearity (which analogizes to "overlapping skill sets" in teams of people).

12.1.5 Interaction between Predictors

In multiple regression modeling, an interaction exists between two predictors X_i and X_j if the effect (read: numerical value of the slope) of X_i changes depending on the numerical value of X_j. The definition is symmetric, so you may in the previous sentence reverse the roles of i and j.

The regression notion of an interaction does analogize, however, to teams of people working on a project. In response to someone working hard, a team-mate might decide to work less, or that intensity might drive them to greater effort as well.

A semantics warning is useful here. In everyday English, an interaction between two people implies some sort of relationship between them. Applying that notion by analogy to two predictors in a multiple regression model suggests that an interaction between two predictors has something to do with their relationship with each other (i.e. with their correlation). That reasoning is incorrect. An interaction may or may not occur between two predictors regardless of whether they are correlated.

12.1.6 Instructive Breakdown of the Metaphor

Quite often, points of breakdown in an analogy are just that: breakdowns. We have identified two points of breakdown here; there may be others, but these two points are, in fact instructive in that they add to the understanding of regression.

We wrote above about the effect of adding people to teams and predictors to models, noting that there is a law of diminishing returns: at some point, adding more predictors to a regression model or people to a team does not improve the product. In fact, in the teams of people case, we all know that adding the wrong person to a team can ruin an entire project. That doesn't happen in multiple regression. The coefficient of determination will always go up when you add one more predictor. At some point, such an addition yields inconsequential gains in R^2, and the researcher may decide that the gain is not worth the price of additional model complexity.

Another point of breakdown is the following. When putting together a team of people to work on a particular project, you usually don't have the luxury of trying various combinations of team-mates. You choose the team based on whatever information you have, then hope for the best. In multiple regression, by contrast, you *can* try many models (teams of predictors) and choose the one that works best for you.

And our second essay. Sometimes, less is more…

12.2 Don't Let the Statistical Tale Wag the Biological Dog

The last sentence in the preface of this book cautions you not to let the statistical tale wag the biological dog. It was our way of saying that we prefer methods that are simpler which

1. Expose the structure of the data, making necessary assumptions clearer and easier to check and
2. Put the biology[2] up front and center, instead of risking it getting lost behind unduly complicated methods.

For us, statistics as a discipline comes to its fullest fruition when it serves the science to which it is being applied. You should consider giving some weight to the interpretability and meaningfulness of a model. Without it, the model is just a mathematical equation. With it, the model may enhance our understanding of some biological phenomenon.

Consider the following... "I used a generalized linear mixed model with a log-link, and the results were significant ($p = 0.001$)."

Question, dear reader: how did this researcher analyze their data? If you claim to have no idea (except that, perhaps, logarithms are involved), then you get full points. That is the correct answer. There are so many designs that could culminate in this description that the description itself has almost no meaning. It leaves the reader or listener in the dark (and possibly the researcher also).

As a "for instance", here is one situation that could land on that description. The researcher has a paired design with, say, 40 pairs. The data values in each group are skewed, so they log-transformed the values. A pause to unpack this before we move on. First, if they are actually interested in the mean of within-pair differences, a sample of size 40 could be sufficient for them to lean on the Central Limit Theorem and assume that the sample mean of differences is approximately Normal. In that case, a one-sample t is appropriate. Having log-transformed, the difference in means on the log-scale back transforms to a ratio of the medians. If that is what they are interested in, fine. But if relative change is indeed of interest, there are ways to address that directly with paired data; no need to log-transform. See Chapter 15 for that discussion.

A paired data analysis is an example of a simple block design. The pairs constitute the blocks, and the entire set of treatments (two only) appears in each block. So, they could call their design a block design, a step up in spiffiness from the simpler "paired design." In a block design, the blocks can be called a random effect, since there is no interest in comparing this block to that. The other factor (the treatment of interest) can be called a fixed effect. If you have a design with a fixed effect and a random effect, you can now declare that you have a mixed model. Now THAT is fancy. Oh, wait. Technically, this model (and most that are usually used by researchers) fall into the mathematical family of "linear models." In Chapter 8, we explained that this is an allusion to a mathematical construct called a "linear combination of the parameters." Fancier and fancier.

Oh, wait. They log-transformed their data. One fancy leap for fanciness, a very small step for mankind: a generalized linear mixed model with a log-link. It's not wrong to say that or use that. But

1. Such nomenclature is not very transparent (as we showed here, it does not clearly say what the design was), and
2. That lack of transparency means that the conditions under which the analysis is valid are not clear either.

The validity conditions are precisely those for the paired t, namely that the sample size is large enough to justify leaning on the Central Limit Theorem to argue for approximate Normality of the mean of the differences.

Paul Murtaugh (2007) published a paper pleading for simplicity in ecological data analysis. He was sparked in part by his work as a statistician at the University of Oregon, and a quote from one of his clients leads off the paper: *I can easily test the hypotheses by simple t-tests but want something more "elegant" that will fit well with a "better" journal.* Paul goes on to use one of his own studies to illustrate his point. If you read the initial description of the analysis in his study (a legitimate analysis, to be sure), you might be as baffled as the statement that starts this section. Subsequently, he uses a simpler approach, which we will introduce as a response feature analysis (Chapter 17). This essay is an amplification of Paul's plea. Don't let the statistical tale wag the biological dog.

12.3 Chapter Summary

Ken has used the metaphor of people working on group projects many times since it first popped into his head, with good results. Students can think about relatively technical issues in multiple regression with less intimidation on their part, and without needing to base the discussion in the language of matrix algebra. The math cannot be avoided altogether, but the use of this "teams of people" metaphor can get you quite far down the road to successful understanding, relatively math-free.

The goal of doing statistical analyses should be to shed light on the underlying processes, be they biological or otherwise. Your choices of methods and nomenclature will help brighten that light or dim it.

Notes

1 To keep complexity down to a dull roar, we did not consider any interactions.
2 We use the term as a placeholder here. Your data needn't, of course, be biological.

Reference

Murtaugh, P.A. 2007. Simplicity and complexity in ecological data analysis. Ecology 88(1): 56–62.

13

Introduction to Logistic Regression

We studied in the multiple regression chapters how to handle categorical predictors using (0, 1) indicator variables. Logistic regression is a regression method specialized for the case where the response variable is categorical. There are versions of logistic regression for cases where there are more than two categories (nominal logistic regression) and even ordered categories. We will consider only binary logistic regression, as it is the foundational and most-used version. Here, there are only two categories, for instance, presence/absence, success/failure, and the like.

Logistic regression does not quite fit into a list of "essentials"; indeed, one of the reviewers of our manuscript noted his surprise at seeing the chapter. Fair enough. This chapter is in no way a full-on working introduction to logistic regression, but there are a few things in particular that might be useful to you, especially since you will undoubtedly run into this method in research articles or reports.

We need to learn about the equivalence of odds and probabilities and learn that the model-fitting works with the log-transformed odds, which is called the logit scale. Inference on the effect of a predictor happens on the odds scale and is multiplicative, captured by the so-called odds ratio. It appeals to intuition that one uses logistic regression to estimate the probability of an event occurring. It is therefore counterintuitive to learn that doing so is sometimes invalid, although you can still estimate the effect of the predictor. You need to learn to differentiate between prospective (legitimate to estimate probabilities) and retrospective (not legitimate) studies.

One important notion to get a grip on at the outset is the equivalence between odds and probability; they are the same thing, differently stated. Logistic regression is amenable to multiple predictors, as well, which we will illustrate with data on the survival of members of the so-called Donner party, using age (in decades[1]) and sex as predictors. Testing the regression assumptions is a bit different for logistic regression. We will show you the how and why of that.

13.1 The Logistic Regression Model (Visually)

In the mid-1840s, a party of California-bound American settlers got caught by winter snows in the Sierra Nevada Mountains. Of the 87 members of the infamous "Donner Party," 39 perished (Grayson, 1990). We will use age and

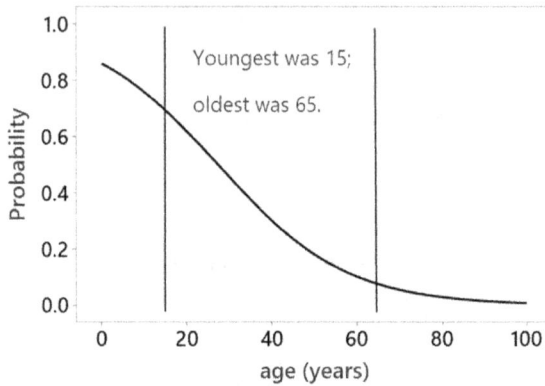

FIGURE 13.1
Probability of survival for members of the Donner party as a function of age.

sex information as predictors of survival. If you are unfamiliar with this sad tale, you can read all about it online.

We will eventually fit the model and interpret the appropriate output, but for now, let's stay at a conceptual level. Figure 13.1 shows how the probability of survival changes with age. The ages of people in the data set ranged from 15 to 65; we stretched the age axis for illustration.

A model for predicting probability must yield values that stay between 0 and 1 (or between 0% and 100% if you prefer). You can see this regression line curving to do just that. The necessary flattening out near 0 and 1 creates certain interpretational problems.

We can see in the graph that the probability of survival declines with age; for a 15-year-old, it is approximately 70% and drops to 10% by age 65. Here we can see that the effect of age changes depending on what age you start at. The change from 15 to 25, for instance, is more dramatic than it is from 55 to 65.

To resolve this problem, we need to take advantage of the fact that probabilities and odds are different ways of expressing the same thing. Let's explore that for a bit before we return to this example.

13.2 Odds and Probabilities: Same Thing on Different Scales

There are three different scales of measurement we need to wrap our heads around. The probability scale is the one we most intuitively look to; it is useful for visual depictions, as shown above. However, as we also saw above, the effect of age, as measured on the probability scale, keeps changing, depending on age itself. If we can shift from probability to odds, we can in fact talk

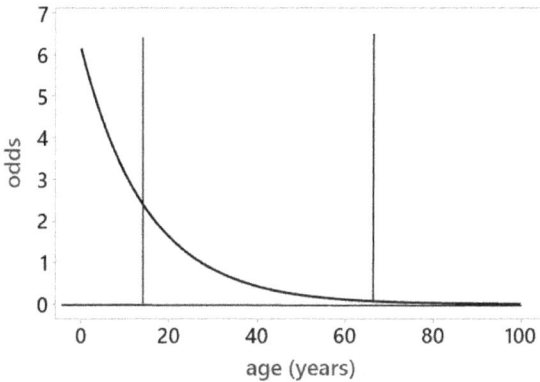

FIGURE 13.2
Odds of survival as a function of age.

about a consistent change in the odds for a change in age. Odds and probability are simply different expressions of the same phenomenon. Given the probability (expressed as a number between 0 and 1) of an event occurring, the odds are $\frac{\text{probability}}{1\text{-probability}}$. Another way to express this is $\frac{\text{prob. of occurring}}{\text{prob. of not occurring}}$.

For example, if the probability of occurrence of some event is 0.8, the odds are 0.8/0.2 = 4. A probability of 0.6 yields odds of 0.6/0.4 = 1.5. Notice that our expression of odds is different than that in common parlance. In everyday English, we would say 4:1 (read as "four to one) for the former, and 3:2 ("three to two") for the latter. In our statistical use, we just do the math and stop[2].

Turning the problem around, given the odds, the probability is $\frac{\text{odds}}{\text{odds}+1}$. For example, "odds of 1.5" is the same as saying the probability is 1.5/2.5 = 0.6.

Back to the logistic regression modeling of survival in the Donner party... Let's examine the odds scale (Figure 13.2), the logistic regression model predicting the probability of survival as a function of age.

There is a direct interpretation here in terms of relative change in the odds. If you look closely, you can see that the odds of survival are approximately cut in half for each advancing decade of age. The odds are approximately 2 for a 15-year-old; they are around 1 for a 25-year-old. It is on the odds scale that we can make sense of the effect of the predictor.

13.3 Modeling Details: We Need Yet One More Scale

Fitting the model requires yet one more change of scale for the predictor. We've moved from probability to odds; now we need to use the natural logarithm of the odds, also called the logit. This scale is shown in Figure 13.3.

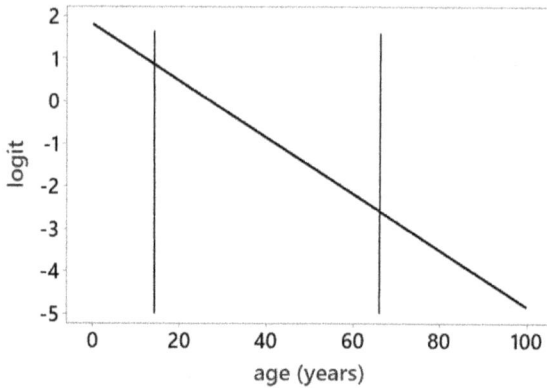

FIGURE 13.3
Log(odds) of survival as a function of age.

On this scale, we see a classical-looking simple linear regression, which can be characterized by a slope and an intercept. This could lead to classically very simple interpretations of the effect of age: we all know how to interpret a slope coefficient in a simple linear regression, except... **what the heck is a logit?** This scale is important, though: it is the scale used by a statistics package to fit the regression.

13.4 Checking the Assumptions

Recall the assumptions behind a classical regression model: Goodness of Fit, homoscedasticity in the residuals, and that the distribution of the residuals is Normal enough. Here, the response is binary. The behavior of variation in Binomial data is well known, and so your statistics package will automatically accommodate that. You do need a sufficient sample size so that the distribution of the parameter estimates is approximately Normal. Some authors would say that you should have at least 20 observations for every variable in the model; others are more liberal. That leaves Goodness of Fit. There are three that are commonly offered, often all three. There is one often called simply a "deviance" test, the Pearson test, and the Hosmer-Lemeshow test. The first two **only** apply to models with **only** categorical predictors. If your model has at least one numerical predictor, you should turn to the Hosmer-Lemeshow test. All three begin with the null hypothesis that the data are adequately modeled; a small p-value suggests that changes should be considered. Perhaps a quadratic is called for, or interactions to be added, or maybe a predictor needs to be log-transformed. We note here that in our examples, the H-L test suggested satisfactory GoF.

13.5 Tying It All Together, with Examples

To sum up, there are three scales in play whenever you do a logistic regression:

1. Probability. Possibly useful for graphics, but see section 13.6.
2. Odds. The scale on which we make inferences on the relationship, via the odds ratio.
3. Logit. This is the scale on which your statistics package estimates the relationship.

We compressed this discussion into a convenient table (Table 13.1); viewing it may help you to become more comfortable with these scales.

In the following examples, we will show you some of the typical computer output and focus on interpretation. Then follows a section on Goodness of Fit, and an important discussion of prospective and retrospective designs.

Example 13.1. Are Yellowstone Cutthroat Trout More Likely to Occur in the Presence of Brook Trout?

In the first year of a study aimed at determining where populations of Yellowstone Cutthroat Trout (YCT) might occur, Wyoming Game and Fisheries biologists surveyed 24 100m stream reaches in the Bighorn Mountains of Wyoming (Dave Zafft, personal communication). Presence (1) and absence (0) of YCT were recorded, as well as a collection of potential predictors, among them the biomass of Brook Trout (BT). Typically, fisheries biologists measure biomass in kg/ha. Assuming a typical stream width of 2.5m, a hectare of surface corresponds to 4km of stream; hence, our feeling that 1kg more of fish per hectare was "insensible." Hence, we rescaled that data into kg/100m of stream length on the grounds that.

TABLE 13.1

Relationships between Probability, Odds, and Logit Scales

Scale	Model
Probability	$\dfrac{\exp(\beta_0 + \beta_1 X)}{1 + \exp(\beta_0 + \beta_1 X)}$
	$\Uparrow \ldots \dfrac{odds}{1 + odds} \quad \Downarrow \ldots \dfrac{prob}{1 - prob}$
Odds	$\exp(\beta_0 + \beta_1 X)$
	$\Uparrow \ldots e^{logit} \quad \Downarrow \ldots \ln(odds)$
Logit	$\beta_0 + \beta_1 X$

The fitted model, using YCT for the presence of YCT and BT for BT abundance, was logit $(YCT) = -1.59 + 0.08 \times BT$. The model was significant at $p = 0.031$, and the resulting odds ratio was $e^{0.04} = 1.04$, with a 95% CI of (1.004, 1.08). Given that the biologists predicted a positive relationship, if any, this could have been done with a one-tailed test ($p = 0.015$). For each additional kilogram of BT per 100m of stream, the odds of finding a cutthroat go up by 1.04 (i.e. they are 4% higher).

Example 13.2. We will now illustrate a multiple regression model for logistic data, using both age and sex for the Donner data. Formally, the regression model we are entertaining is $L = \beta_0 + \beta_1 F + \beta_2 A + \beta_3 (A \times F)$, where L is ln $(odds)$, A is age (in decades), and F is an indicator variable for sex, coded as 0 for males and 1 for females. The model includes an interaction term between age and sex to allow us to consider whether the effect of age is different for females than males.

Notes:

1. The logarithm of the odds goes by the name "logit,"
2. β_0 is the intercept coefficient, and β_i; $i = 1,2,3$, are the respective slope coefficients.
3. If the model is to have an interaction between A and F, then both A and F should be retained in the model also.[3]

Let's examine the model both with and without the interaction term, to better understand its role. Second, we will formalize the test of significance for that term.

Recall that an interaction between F and A implies that one cannot correctly address the effect of F or A without referencing some chosen level of the other. Interpretation of the interaction in this model is simplified by the fact that one of the predictors (F) can only be a 0 or a 1. The interaction term is not significant at alpha $= 0.05$ ($p = 0.086$) but is close. For the sake of discussion, let's suppose we keep it in the model. The full model is $L = 0.32 + 6.92F - 0.325A - 1.616F \times A$. To study this further, let's set F first to 0, then to 1. If $F = 0$, the model reduces to $L(M) = 0.32 - 0.32A$. Now, $e^{-0.32} = 0.73$. Subtracting from one and rescaling to a percentage, we conclude that for every advancing decade of age, the odds of survival for males are reduced by 27%. If F is set to 1, then we get $L(F) = (0.32 + 6.92) - (0.32 + 1.62)A = 7.25 - 1.94A$. Now we get an odds ratio of $e^{-1.94} = 0.14$. The odds of survival for females are reduced by 84% for each advancing decade of age.

Figure 13.4 illustrates these fits for the logit and probability scales.

Given that the interaction term is not quite significant, it is possible that the apparent difference in the slopes for males and females is a random artifact of the data. If we want to remove the interaction term, we must run the model again without it, in which case we get $L = 1.63 + 1.6F - 0.78A$. Notice that the coefficients are all different from the model that includes an interaction term. This is to be expected: any time you remove or add a predictor, all the other terms will get reevaluated.

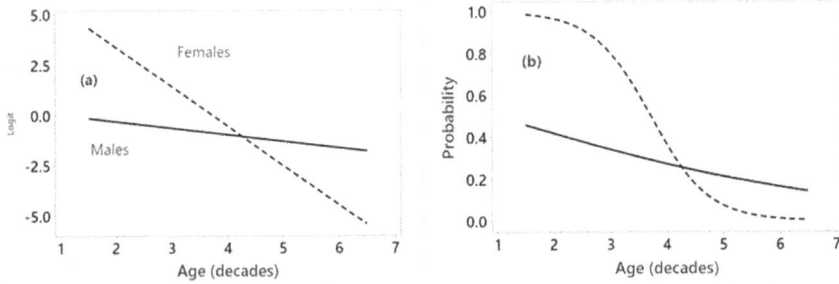

FIGURE 13.4
Logistic regressions for males and females on the logit scale (a) and probability scale (b).

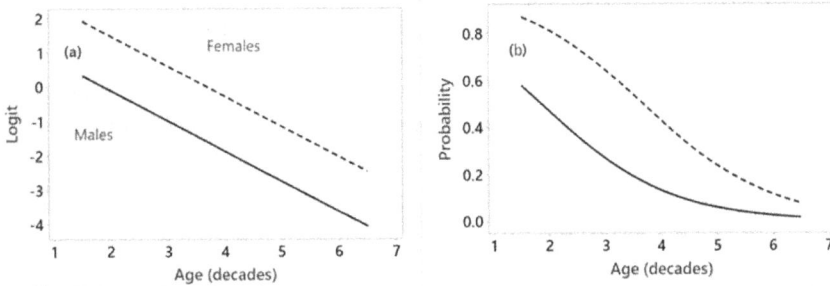

FIGURE 13.5
Logistic regression equations for males and females with the interaction between age and sex. The logit scale (a); the probability scale (b).

Survival goes down with age, marginally significant at $p = 0.036$. $e^{-0.78} = 0.46$. For every advancing decade of age, the odds of survival fall by 54%. The effect of sex is also significant ($p = 0.034$). The odds of females surviving are $e^{1.6} = 4.95$, approximately five times higher than a male, approximately true across age. Graphs shown in Figure 13.5 illustrate these equations for the logit and probability scales.

Which model is best? The interaction term could be included or not. The model with the interaction is more complicated (more interesting?); the other is simpler. This is a question for Tyler, if you can find him.

13.6 Retrospective and Prospective Models (with a Big Take-Home Message)

Logistic regression first became well-used and studied in the context of medical studies, with the outcome being, for example, whether a person has some condition of interest. In that context, two main types of designs arose.

In one design, the researcher chooses some number of subjects who display the condition of interest (so-called "cases"), and another group who do not (the "controls"). Certain predictors are measured on each subject. These case/control designs also became known as *retrospective* since the event of interest has already happened.

In the other common design, subjects are chosen and followed over time, during which some of them will develop the condition of interest. These studies became known as *prospective* studies.

In a prospective study, the data allow a natural estimate of the underlying probability of the occurrence of the event; the predictors show how they might affect that probability. In a retrospective study, however, that is not possible. For instance, if you happen to choose equal sample sizes for the case and the control groups, your data will indicate that the underlying probability of the event is 50%. Fortunately, the slope coefficients are still valid, so estimating the effect of a predictor by examining the odds ratios is still valid.

These terms and their emphasis on time: "did the event already happen or are we waiting to see *if* it happens?" are unfortunately misleading.

Examples

Example 1. Nate Bumpus collected 24 sparrows that had perished during a particularly severe winter storm; in addition, he trapped 35 sparrows that had survived. On each, he recorded their size by the length of the humerus bone (see Manly et al, 2025, Chapter 1, for a more detailed introduction). A logistic regression could be used to estimate the effect on the odds of survival of size, but the study is retrospective since he chose the numbers in each group. Probability estimates are invalid.

Example 2. Researchers measured the size (diameter four feet above ground) of many whitebark pine trees and noted whether those trees had been infested by pine bark beetles. The event of interest (infestation by beetles) has already happened. BUT... researchers did not choose how many infested and non-infested trees would be in their sample. Effectively, they went *prospecting*[4] into their data to see which trees would show the problem. This is a prospective study. Estimates of probability (or odds) of infestation are legitimate.

Example 3. Fisheries biologists recorded the amount (biomass, in kg) of BT in a sample of 100m stream reaches; in each, they also recorded the presence or absence of Yellowstone Cutthroat trout. The two species are sympatric, meaning they do well in similar environments. The question was whether the abundance of BT would be reflective of the odds of there being cutthroat trout present also. Since they did not predetermine how many of their sample reaches would have cutthroat trout, this is a prospective study. Estimating the probability of the presence of cutthroat trout is legitimate.

13.7 Chapter Summary

Logistic regression is a specialized regression method for the case of having categorical response variables. We studied binary logistic regression, wherein there are only two categories for the response. They may be, for instance, (presence/absence), (success/failure), and the like. A common nomenclature for this setting is to refer to "cases" (observations of the event of interest) and "controls" (the event was not observed).

We showed you the three different scales used in the modeling (and how to transform from one to another):

1. Probability (useful for visuals when appropriate)
2. Odds (the scale used for inference on the effect of a predictor via the odds ratio), and
3. Logit (the scale on which the model is fitted.

The probability and odds scales are equivalent and are often used interchangeably. It is counterintuitive to learn that it is not always legitimate to estimate probabilities, depending on the study design. In so-called retrospective studies (defined by the researcher choosing the samples of "cases" and "controls"), the effect of a predictor can be legitimately estimated, but direct estimates of the probability of the event of interest cannot. A prospective study is defined by the fact that "cases" and "controls" appear by natural chance in the sample. In these studies, it is legitimate to estimate the probability of an event occurring.

Notes

1 We usually record age among humans in years. But the change in odds of surviving for, say, a 34-year old and a 35-year-old is very small. One year is too small a step size for ready discussion.
2 You are unlikely to get into trouble in a presentation if you express the odds in "street style"; in fact, your audience might appreciate it.
3 This is true for any setting where an interaction is being considered. The two (as we call them) "parents" must be kept if their interaction is in the model.

4 This is how we prefer to use the term *prospecting*. If you do not choose ahead of time, how many events/nonevents you will get, that study is a prospective one, and probability estimates are legitimate.

References

Grayson, D.K. 1990. Donner party Deaths: a demographic assessment. Journal of Anthropological Research 46(3): 223–242.

Manly, B.F.J., J.A. Navarro Alberto and K. Gerow. 2025. Multivariate Statistical Methods: A Primer. Chapman and Hall. 279 pages.

Part IV

Methods for Comparing Groups and an Approach to Repeated Measures

You might have been surprised to see the chapters on t-tests for differences in means and one- and two-sample procedures for Binomial proportions placed closer to the end of a text rather than the beginning. To be sure, these methods are among the simplest, and so there is a good reason to start a course (or book) with those topics. Over the years, seeing the somewhat lukewarm interest of students prompted Ken to switch it up. Most analyses that most people do most of the time involve regression models or ANOVA, often with multiple predictors. Starting with those topics has helped with initial engagement. Of course, in a book, you are free to read the chapters in any order you wish.

In addition to the predictable material on procedures for means and proportions for one and two samples, including the t-tools for paired data, we extend the material in the following ways that we hope you find useful.

- The classical statistics toolkit takes aim at differences in means, calculated by subtraction. Sometimes, though, relative change rather than an increment of change is of interest. The freedom to make that choice is not often highlighted in one or two semesters of coursework; we devote a chapter to the subject to expand your inferential choices.

- ANOVA, meant for comparing means from multiple groups, is a natural extension of the two-sample procedures, so there is a chapter

DOI: 10.1201/9781003609605-17

on it also. We will illustrate the fact that the so-called ANOVA procedures and regression modeling are in fact just superficially different approaches the same problem. We will show you how to approach so-called contrasts among treatment groups and give you our take on the problem of multiple comparisons.

- It might seem odd to have a chapter on repeated measures analyses in a book called "Essentials…" The approach we espouse to repeated measures data (it is not our invention) enables you to analyze such data with the statistical tools that you are already familiar with. The approach has two steps. First, choose a summary measure that answers a question of interest. Apply that to each subject (person, plot, whatever). Then do a statistical analysis of the summary measures. Our examples will illustrate that the approach is flexible and scientifically powerful in that it allows you to answer quite more interesting questions than a classical repeated measures approach can easily do.

14

One- and Two-Sample Methods
for Means and Proportions

This chapter shares mostly standard material, focusing on means: estimation of and testing for a mean from a single sample, the difference in two means from independent samples, and the mean of differences for paired data. That is followed by a section on inference for Binomial proportions from single samples and differences in proportions for two independent samples.

Most research deploys analyses that are more complicated than these, but these methods are perhaps foundational in that they lay the groundwork for the other methods.

What's novel? We will claim our approach to the equal variances question is novel, arguing that presuming equal variances is often a logical consequence of the null hypothesis, and that testing for that equality almost rises to the level of being in and of itself a circumstantial evidence test of the usual null.

14.1 Use of the One-Sample t-Tool

We used single-sample data for a measured variable, with the goal to estimate a population mean, to illustrate the Central Limit Theorem (CLT) and introduce standard errors (Chapter 1) and to introduce hypothesis testing and confidence intervals (CIs; Chapter 2). So, here we will quickly summarize the tools, and then focus on the assumptions where, we think, things get more interesting.

- The parameter of interest is a population mean μ, estimated by the sample mean $\bar{y}$.
- The SE of the sample mean is estimated by $s/\sqrt{n}$, where s is the sample SD, and n is the sample size.
- Provided conditions are met, inference is done using a t distribution with n-1 degrees of freedom.
- A test against some mean[1] μ_0 compares the resulting p-value against your chosen alpha level; it is common practice in science to use $\alpha = 0.05$.

DOI: 10.1201/9781003609605-18

- A CI for μ takes the form $\bar{y} \pm t_{n-1,CL} \times SE(\bar{y})$, where the value of the t multiplier depends on the choice of confidence level. For the conventional 95% level, $t \approx 2$ for all but very small sample sizes.

14.1.1 Assumptions

As is the case for any statistical tool, there are validity conditions to be met. For the one-sample setting, the t-tools are valid if

1. The sample is a random[2] sample of independent[3] observations.
2. The distribution of Y is Normal enough.

There are a few things to unpack in point (2). First, why are we talking about Y, the random variable itself? As you might recall from the regression modeling chapters, we ask about the distribution of the residuals; the regression assumptions say nothing about Y. Second, why did we say, "Normal enough"? Shouldn't the assumption be that Y has a Normal distribution?

In fact, the assumption *is* about the residuals, but in this case, we can cheat and just look at Y. Residuals in a model are the difference between individual values of Y and the fitted, or estimated, values. In a regression model, the fitted values come from a model $b_0 + b_1 X$, where the three terms are the intercept, slope, and the predictor, respectively. In that case, each value of the response $(Y_i; i = 1, 2, ..., n)$ has an associated residual: $res_i = y_i - b_0 + b_1 x_i$. Here, though, the model is much simpler: the sample mean $\bar{y}$ is used to estimate the population mean μ. Thus, the residuals are simply $res_i = y_i - \bar{y}$. Let's see what that means for the geyser waiting times data we used to illustrate the CLT in Chapter 1. The bin endpoints chosen by the statistics package are slightly different for the two histograms, such that panels (a) and (b) Figure 14.1 are not perfectly identical, but a description of the shape of the distributions (two bumps; a smaller one to the left and a larger one to the right) is indeed the same.

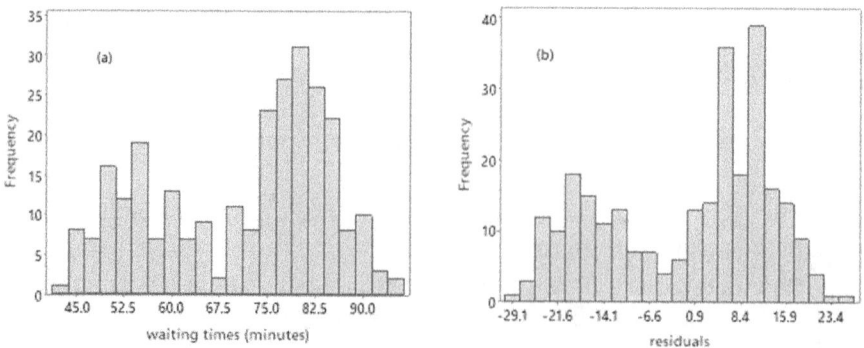

FIGURE 14.1
Geyser waiting times (a) and residuals (deviations from the mean) (b).

The information we require regarding the shape of the distribution of the residuals can be had by simply examining the shape of the distribution of the original data. We needn't take the steps to calculate the residuals; it suffices to look at Y itself. Consider the values of Y and residuals for the geyser waiting times (Figure 14.1).

These data are clearly not Normally distributed. Despite that, as we showed in Chapter 2, the sample size of 272 is easily large enough to declare that the distribution of the mean is approximately Normal. This in turn justifies the use of the t distribution for inference.

14.1.2 Is the One-Sample *t*-Tool Legitimate for *n* = 7?

Suppose you had a smaller sample size? Suppose it was only 7. The relevant matter is the distribution of values in the population that gave rise to your sample. Ideally, we would examine the distribution of our sample to see what we can learn about the distribution of the population. Figure 14.2 shows six examples of samples of size *n* = 7 from mystery distributions. Can you guess what distributions they come from?

The first row are random samples from a Normal distribution; the second from a Negative Binomial, which is skewed right. Full disclosure: we cherry-picked these examples from 15 in each case, but these aren't the most or only

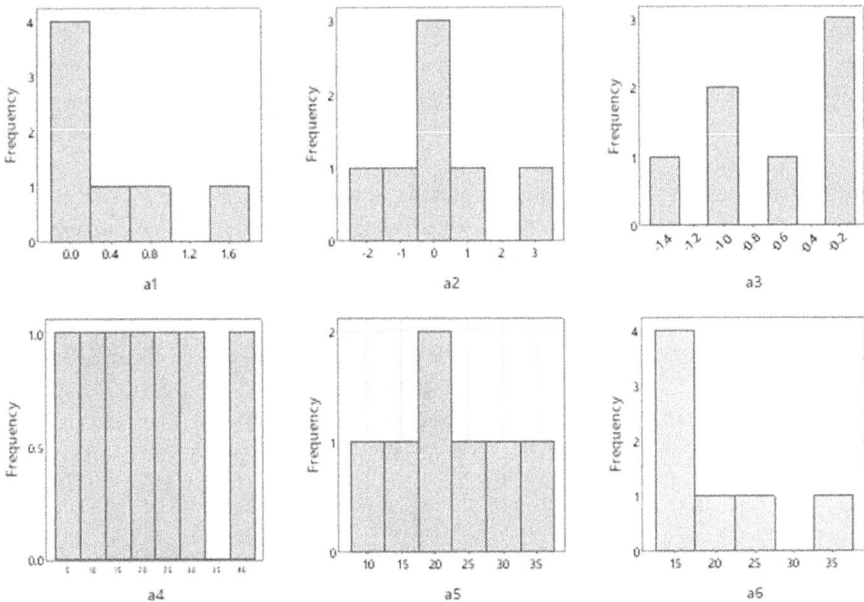

FIGURE 14.2
Random samples of size *n* = 7 from mystery distributions.

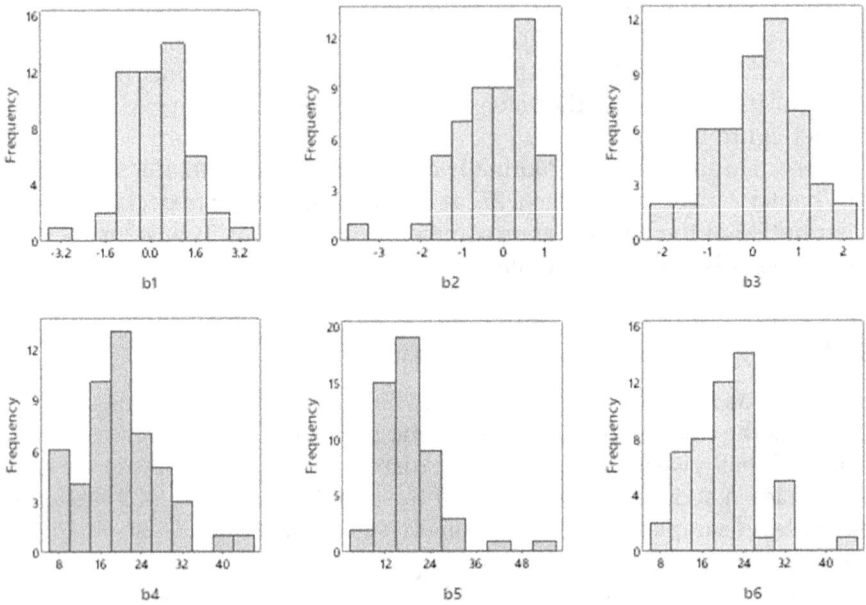

FIGURE 14.3
Random sample of size $n = 50$ from a Normal distribution (top row) and a Negative Binomial distribution (bottom row).

misleading cases from the fifteen. And, true, a sample of size seven is quite small, but not unheard of. Our point here is that you cannot trust a small sample to represent the distribution of the population whence it came. So, at risk of overstating things, testing for Normality is a waste of time for small samples.

On the other hand, Figure 14.3 shows six samples of size $n = 50$. The top row graphs are from the same Normal distribution as the top row of Figure 14.2. The bottom row ones come from the same Negative Binomial distribution as those in the bottom row of Figure 14.2.

Even with $n = 50$, the samples can't be counted to look precisely like the distributions they came from, but they are way better at it than are samples of size $n = 7$. For a large enough sample size, we don't particularly care about the distribution whence the data came, because the CLT assures us that the distribution of the mean will be at least approximately Normal, which makes the use of the t distribution valid. So, for large sample sizes, testing for Normality is a waste of time.

It is of course a misleading stretch to flat-out say that testing for Normality is a waste of time, but we *are* trying to point out that it needs to be done thoughtfully. For small samples, you might need to argue from data other than your own[4] that the distribution is close enough to Normal, and for large samples, you need to learn to trust the CLT.

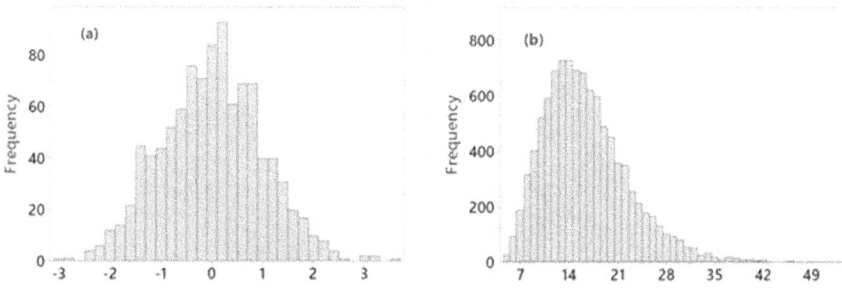

FIGURE 14.4
Samples of size $n = 1000$ from the Normal (a) and Negative Binomial (b) distributions used above.

So that you can see the shapes of the distributions used above, we show in Figure 14.4 samples of size $n = 1000$ from those two distributions.

14.2 Differences between Means from Two Independent Samples

After a particularly severe winter storm in 1898, Hermon Bumpus collected measurements of house sparrow size, as measured by humerus length, for males[5] that perished in the storm ($n = 24$) and some that survived ($n = 35$); data are displayed in Figure 14.5. Here, we are interested in the question, "Did the survivors differ in size from those that perished?"

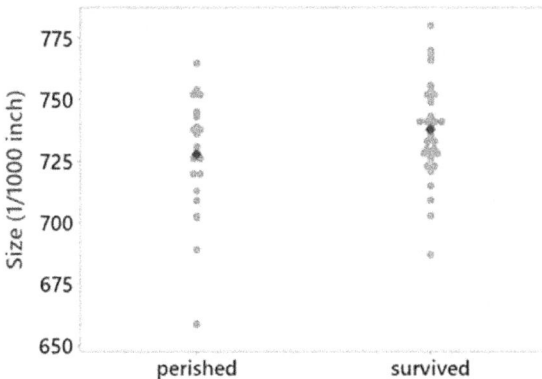

FIGURE 14.5
Individual value plot of sparrow lengths. Means are indicated by black diamonds.

TABLE 14.1

A Brief Numerical Summary of the Sparrow Data

Variable	n	Mean	SE(M)	SD	Min.	Q1	Median	Q3	Max.
Perished	24	727.9	4.81	23.54	659	714.75	733.5	743.75	765
Survived	35	738.0	3.35	19.84	687	728.00	736.0	752.00	780

A common study design to examine the effect of a particular phenomenon (a certain treatment, or perhaps a given circumstance) is to have two
independent samples, one associated with the "treatment"; the other is often
called the control group. Here, those particular terms don't apply; instead, we
have those that perished and those that survived.

We will assume here that you are already comfortable with CI properties
and the basics of hypothesis testing, as well as the role of the CLT in justify-
ing the use of the t distribution.

We will use the data briefly summarized in Table 14.1 to illustrate the use
of a two-sample t for estimation, via CIs, and testing. Here, interest is in the
parameter[6] $D = \mu_S - \mu_P$, as estimated by $\hat{D} = \bar{y}_S - \bar{y}_P$.

What we need to show you here is straightforward, given that you are comfortable with tests, CIs, and the CLT, except for one thing. You need to decide
whether to assume equal variances or not. That choice has implications for
the degrees of freedom of the t-distribution and the formula for the standard
error. So, let's get that out of the way. Before we proceed, there is something
you need to know and something you might need to be reminded of.

True fact: If you combine two independent statistics into one, by adding
them or subtracting one from the other, the variance of that combined statistic is equal to the sum of the two individual variances.

Reminder: The SE of the mean from a random sample is estimated by $s/\sqrt{n}$.
The square of this is called the variance of the mean: s^2/n. You will see that
latter form appear below because the first step to calculating the SE of a statistic is to calculate the variance. Then take the square root.

Suppose you assume the variances are equal. In that case, each sample
provides an independent estimate of the presumed common variance, and
the best thing to do would be to combine them into a single pooled estimate:
$s_p^2 = \frac{(n_1-1)s_1^2 + (n_2-1)s_2^2}{(n_1-1)+(n_2-1)}$. This is a weighted average of the two, with the estimate from
the larger sample getting more weight. Then the standard error of $\hat{D} = \bar{y}_1 - \bar{y}_2$
is $SE(\hat{D}) = \sqrt{\frac{s_p^2 + s_p^2}{n_1 + n_2}} = s_p\sqrt{\frac{1}{n_1} + \frac{1}{n_2}}$. In this case, the degrees of freedom for using
the t distribution are calculated by $n_1 + n_2 - 2$.

Suppose you *don't* assume they are equal. Then the standard error of
$\hat{D} = \bar{y}_1 - \bar{y}_2$ is $SE(\hat{D}) = \sqrt{\frac{s_1^2}{n_1} + \frac{s_2^2}{n_2}}$. The degrees of freedom formula is more complicated[7] and generally yields a value less than $n_1 + n_2 - 2$.

Since the two approaches use different arithmetic, they won't yield the
same answers (p-values and CIs), although they often don't differ by much,

especially if the sample sizes are equal. Still, they aren't the same, so you need to choose which way to go.

What do we recommend? For estimating the difference between the two means, we prefer not to assume equal variances. The phrase "the difference" asserts that there *is* a difference in the population means. If so, then the two samples come from different populations, each with its own variances. True, the two variances might be similar, but we see no basis to argue that they must be equal.

Our view on the matter of testing, however, is different. A criminal court case starts with the presumption of innocence for the stated crime and then examines the evidence, considering that presumption. Similarly, a hypothesis test for the difference in two means against the research hypothesis that the means do differ starts with the presumption that there is one mean, in which case the two individual sample means both estimate it. A statistical hypothesis test focuses on some parameter, an individual value. That said, the underlying scientific null is often broader: that the two groups do not differ, or that, in the case of a control/treatment setting, the treatment made no difference. If that is so, then the two samples come from a single population, in which case not only is there but one mean, there is also but one variance. So we argue that the presumption of equal variances is consistent with the null hypothesis.

Indeed, it is commonplace with ratio data (values are zero and above, and which are ubiquitous in scientific studies), that variation in a variable is positively correlated with the mean: larger means come along with larger variances. So, if you test the variances (as some would suggest) and find them to be different, that might be circumstantial evidence against the null hypothesis of equal means.

To sum up:

1. For CIs, we do not assume equal variances.
2. For testing, we do.

That said, it is certainly the case that there is no consensus on the matter (remember Tyler Johnson?). Views on the matter vary widely; we document some of them in an Endnote to this chapter.

Now that we have cleared the air on that one matter, here is some of the math behind the two-sample *t*-tool. You will of course have a statistics package doing this for you, but it is useful to have some idea of what is under the hood. First, we unpack the construction of a 95% CI, using the sparrow data as an example.

$$\hat{D} \pm t_{43,0.95} \times SE\left(\hat{D}\right) = 10.08 \pm 2.02 \times 5.85.$$

where

1. $\hat{D} = \bar{y}_S - \bar{y}_P$ symbolizes the estimated difference in means.

2. $t_{43,0.95}$ represents the value from a t distribution with 43 degrees of freedom such that $\pm t$ captures the middle 95% of the distribution, and

3. $SE(\hat{D}) = \sqrt{(SE(\bar{y}_S))^2 + (SE(\bar{y}_P))^2}$ is the estimated standard error of the difference, the formula for each element is the formula is $SE(\bar{y}) = s/\sqrt{n}$, where s is the sample standard deviation (SD; it estimates the population SD, often denoted by σ). Here, $s_p = 23.5, n_p = 24, s_S = 19.8,$ and $n_S = 35$.

For these data, 95% CI for the difference is (–1.73, 21.9).

In our sparrow example, the question was posed as: do the two types of sparrows (survivors and casualties) differ in size, as measured by humerus length? Here, then, we have $H_R : \mu_S - \mu_P \neq 0$, leading to $H_0 : \mu_S - \mu_P = 0$. The resulting p-value is 0.081, larger than the conventional $\alpha = 0.05$, and so we would fail to reject the null hypothesis. We did the test using the "equal variances" condition because asserting equal variances is consistent with the null hypothesis. In this case, we might be able to argue, as a research hypothesis, that larger birds might have a better chance of withstanding the hardship of a severe winter storm. In that case, we have $H_R : \mu_S - \mu_P > 0$, leading to $H_0 : \mu_S - \mu_P \leq 0$. Since the actual difference (10.08) is consistent with the alternate hypothesis, we can simply take the p-value from the two-sided test and halve it: $p = 0.04$. This is modestly significant against the null, at $\alpha = 0.05$.

14.3 Differences in Means with Paired Data

In a study on how schizophrenia physically alters the brain, the hippocampus[8] size was measured in each twin in fifteen sets. In each set, one suffered from schizophrenia; the other did not. It was suspected that those suffering would have smaller hippocampi, since one of the signs of the condition is loss of long-term memory. Let's start with an estimate of differences in average size, if any, and then a test.

A paired t analysis is simply a one-sample t applied to pairwise differences. Here, since the sample mean of sizes of unaffected (U) hippocampi is larger than that of the affected (A) sizes, we will do the subtraction as U – A, to yield a positive number for the difference.[9] The sample size of 15 is modest, but the fact of subtracting one value from another within each pair creates data that are at least slightly closer to a Normal distribution than were the original values. Still, some caution in interpretation would be prudent.

The mean of the differences is 0.2, the SE is 0.06. A 95% CI is (0.07, 0.33). These are small numbers, but keep in mind that they are an artifact of measurement units. Since a directional hypothesis was suggested, a one-sided interval, a so-called lower bound (LB) would be appropriate. Here, LB is 0.09. We can be 95% confident that the true difference is at least 0.09. A two-sided test yielded a p-value of 0.006, stout evidence that the difference is real. The analogous one-tailed test has a p-value of 0.003. There is nothing overly novel to see here, as far as the analysis itself goes.

What might be of interest, though, is to demonstrate the power of pairing, where it is appropriate. Figure 14.6 shows the raw data, with lines joining the values within pairs. In all but one case, the hippocampus of the affected twin is smaller than that of its unaffected counterpart. So even though the effect seems small, it is quite consistent across pairs.

If you did not know the data were paired or ignored that information, a two-sample t-test produces a p-value of 0.056 for the two-sided test (0.028 for the one-sided test). Further, the classical CI is (-0.01, 0.40). By taking appropriate advantage of the pairing, a much more powerful test and precise estimate ensues.

A paired design is beneficial when there is some basis for selecting pairs of entities such that the measurements of interest are likely to be similar within pairs. The sharper the basis for pairing, the more powerful the result will be. A common example is to measure something on the same experimental unit (a subject or plot, for instance) at two points in time, or under two different conditions. In agricultural animal studies, selecting two animals from the same litter would be a good basis for a paired design. Twin studies in humans are of course another great example.

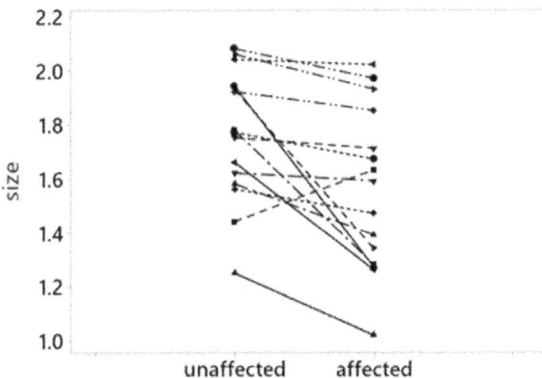

FIGURE 14.6
The hippocampus data, showing the pair affiliation of the values.

14.4 Estimation and Testing for Binomial Proportions

Binomial proportions data arise when each datum is recorded as belonging in one of two categories. Examples of such categories include Present/Absent, Male/Female, Success/Failure. We assume some proportion p in the population falls into the category of interest (which you are free to choose), and the goal of the analyses is to use sample proportions to make inferences to it. Conceptually, the methods here parallel those for one- and two-sample settings for estimating means, with the following principal difference.

We use the standard Normal distribution for inference rather than the t distribution. Essentially, this is because if you know p or have an estimate of it, the SE follows immediately (i.e. it needs only p and n) because (using as data "1" for the category of interest and "0" for the other), the sample SD is known given the estimate for p. We will illustrate below.

14.4.1 Inference from a Single Sample

Is the mortality rate among Bonneville Cutthroat Trout (BCT) in the Thomas Forks irrigation canals in SW Wyoming less than 30%?[10] In the summer of 2002, Amy Schrank (Schrank and Rahel, 2004) tracked 40 radio-tagged BCT. Over the course of the summer, nine (22.5%) died. We will use these data to illustrate CI construction and hypothesis testing for Binomial proportions.

Assuming that the survival of one fish is independent of the survival or death of any other fish, these data are readily modeled by the Binomial distribution. A natural estimator for the true population proportion from the data is the sample proportion $\hat{p} = \frac{Y}{n}$, where Y is the observed number of events[11]. If we record the data as a "1" to denote the death of a fish and "0" to denote survival, the sample proportion is the average of those data values.

One feature of Binomial data is that the standard error of the sample proportion is determined solely by p and n: we estimate it with $SE(\hat{p}) = \frac{\hat{p}(1-\hat{p})}{n}$. Calculate $\hat{p}(1-\hat{p})$ for a variety of values of $\hat{p}$, and you will see that it is at its largest when $\hat{p} = 0.50$ and diminishes as it gets smaller or larger. If something is very likely or very unlikely to occur, then you know what to expect from data: mostly zeros with a few stray ones (or vice versa). In other words, the data set is not very variable. The estimated mortality rate for the BCT is 9/40 = 0.225 (22.5%). The SE of this rate is estimated to be $SE(\hat{p}) = \sqrt{\frac{.225(1-.225)}{40}} = 0.066$ (6.6%).

14.4.2 Validity Conditions for Using the Normal Distribution

Using the z distribution for inference on a proportion is justified if the following conditions are met.

1. The sample should be no larger than 10% of the population (if the sample *is* larger, the sampling distribution of the proportion becomes non-Normal)[12].

2. On the other hand, the sample needs to be large enough that the expected number[13] of 1's and 0's are both larger than 10. For the sake of a label, we'll call this the "10/10" condition[14].

14.4.3 Hypothesis Testing for a Single Binomial Proportion

I assume you are comfortable with the setup: declare your research and null hypotheses and choose alpha. These days, statistics packages usually offer the alternative of doing the calculations using the Binomial distribution directly. We have no strong preference between the two methods. We note, however, that markedly different p-values suggest that the Normal approximation version is not valid, likely due to a small sample size. In that case, the p-value arising from the Binomial distribution is more valid.

In our example, we are interested in seeing if the mortality rate has dropped from the (contrived for this example) historical rate of 30%. That informs the alternate hypothesis[15]: $H_R : p < 0.30$, and the null $H_0 : p \geq 0.30$. If the null hypothesis is correct, then we might expect $0.3 \times 40 = 15$ mortalities and 25 survivors in our sample of 40 fish. This meets the 10/10 rule, so using the Normal distribution ought to be valid. A statistics package will yield the p-value: 0.150. We do not have enough evidence (against alpha = 0.05) with which to reject the null hypothesis. The fact of our observed value being lower than 0.30 appears to have just been random.

14.4.4 Confidence Intervals for a Single Binomial Proportion

I will consider three methods for calculating CIs for a proportion. First, the standard method, which, ironically, is losing its status as the standard. Second, there is a method that uses the Binomial distribution directly, the so-called "exact" method[16]. This is currently the default in many statistics packages. The third method, which we believe will become standard, is called the Agresti-Coull interval after its creators or the "+4" method, after an easy-to-implement approximation, valid for 95% intervals. For our purposes here, we will use 95% as our choice, in part because it is complementary to the alpha (0.05) we used in the hypothesis test.

The standard method uses the formula $\hat{p} \pm z_{CL} \times SE(\hat{p})$, where z_{CL} is the value from the standard normal distribution such that $CL\%$ of the distribution lies between z_{CL} and $-z_{CL}$. For our example, $z_{.95} = 1.96$. For our example, the classically constructed 95% CI is $\hat{p} \pm z_{CL} \times SE(\hat{p}) = 0.225 \pm 1.96 \times 0.066 = 0.225 \pm 0.129 = (0.10, 0.35)$. Note, though, that the validity conditions for using the Normal distribution are slightly shaky, since there were only nine mortalities. Calling the interval "approximately 95%" would be wise.

One alternative is to use the Binomial distribution directly (the so-called "exact" method), available in most statistics packages. Here, the exact CI is (0.11, 0.38). This is slightly shifted and wider than the classical interval.

There is a third alternative that uses slightly different arithmetic for forming the interval. Agresti and Coull (1998) showed that superior CI behavior (i.e. the actual confidence level is closer to the stipulated level) can be had by doing the usual CI calculation on fudged data: for a 95% CI, add two successes and two failures[17], which you can do easily when putting your data into a statistics package. Symbolically, that amounts to defining $p_{AC} := \frac{Y+2}{n+4}$. We emphasize that you are not changing your estimate ($\hat{p} = 0.225$ is still it), it is simply a different algorithm for getting a CI, an algorithm with apparently better properties.

For the BCT data, that would be 11 1s and 33 0s. The resulting 95% CI is (0.122, 0.378). The Agresti-Coull method is becoming widely recommended, as it balances ease of implementation and technical properties. This "+4" trick is valid only if you choose 95% as your confidence level.

14.5 Inference on Two Independent Binomial Proportions

Does being infected with blister rust increase the probability of pine bark beetles infesting whitebark pine? As part of a larger study done by Bockino and Tinker (2012), we randomly selected data from 150 trees, 100 with blister rust (*BR*) and 50 without (*NR*). Of the 100, $\hat{p}_{BR} = 69\%$ were infested by bark beetles; $\hat{p}_{NR} = 42\%$ of the 50 uninfected trees had been hit by bark beetles. Is the difference (69% versus 42%) just chance, or is it evidence that rust-infected trees have lowered resistance to the beetles?

The data are suitable: they easily meet the 10/10 rule for using the Normal distribution for inference, but we must assume that a tree being infected does not have any bearing on neighboring trees being infected. Here, we have a one-tailed test: is there an increase in beetle infestation in association with blister rust? In short, $H_R : p_{BR} - p_{NR} > 0$, and $H_R : p_{BR} - p_{NR} \leq 0$. The observed difference is $\hat{p}_{BR} - \hat{p}_{NR} = 27\%$, which yielded $p = 0.001$, with a[18] 95% CI of (11%, 43%).

An observed difference in proportions of 27% might have different connotations depending on the location of the two proportions. For instance, going from 5% to 32% might be a rather dramatic shift. Here, one proportion is not far below 50%, the other somewhat above. As an alternative (logistic regression in Chapter 13), we could shift scales from probabilities to odds and then form the odds ratio. In this case, $odds_{BR} = \hat{p}_{BR}/(1-\hat{p}_{BR}) = 0.69/0.31 = 2.23$, while $odds_{NR} = \hat{p}_{NR}/(1-\hat{p}_{NR}) = 0.42/0.58 = 0.72$, so we can say the odds of being infested by bark beetles are three times as high ($2.23/0.72 = 3.07$) if the tree has already been infected by blister rust.

14.6 Chapter Summary

We studied details for testing and estimation for one and two-sample settings, for both measured variables and Binomial proportions. The methods involve the use of the t distribution for means of measured variables and the standard Normal (a.k.a. z) distribution for proportions.

There arises in the case of two independent samples whether you should assume equal variances. Feelings about that vary (see our end note on the subject). It is standard practice to assert that they are equal for comparing two proportions, since that follows from the null hypothesis of no difference. We follow that logic for the two-sample t test for means also.

The classical paired-t reduces to an analysis on a single sample, which is the result of subtracting values within pairs. This within-pairs subtraction creates a sample that tends to be less non-Normal than the original values since the CLT works its magic on adding or subtracting values.

Appendix: Two Equal or Not Two Equal (Variances, That is)

What do other authors and teachers recommend? The authors discussed here were selected by Ken, based on the fact that he has their book on his bookshelves, and they are individuals whom he respects for their work as statisticians, and for the contemporary ones, their work as teachers of statistics. So this is a nonrandom, convenience sample. Take it for what it is worth.

Let's begin with our forefathers.

Ronald Aylmer Fisher. Ronald Fisher worked as an agricultural geneticist, working at Rothamsted Experimental Station in the United Kingdom in the 1920s. The inventor of the F test (used in ANOVA and regression modeling), among other things, he is recognized[19] as "a genius who almost singlehandedly created the foundations of modern statistical science" and "the single most important figure in 20th century statistics."[20] The foregoing was cribbed from the Wikipedia entry on Fisher.

In his book, *Statistical Methods for Research Workers* (seventh edition, 1938), Fisher espoused that one should assert and use equal variances for testing the difference between two means. Indeed, he makes clear (page 130) that this is not merely an assumption:

> "It has been repeatedly stated, perhaps through a misreading of the last paragraph, that our method involves the "assumption" that the two variances are equal. This is an incorrect form of statement; the equality of the variances is a necessary part of the hypothesis being tested, namely that the two samples are drawn from the same Normal[21] population."

Fisher did not address the equal variance question in the context of CIs (i.e. estimating the size of a difference). Back in the day, the focus was almost exclusively on testing.

George Snedecor and Bill Cochran. George Snedecor made many contributions to the foundations of statistical methodology; he founded the first academic department of statistics in the United States at Iowa State University. His 1938 text, *Statistical Methods*, became an essential resource.

In the fourth edition of his book (1948), George repeated Fisher's argument in favor of asserting equal variances for testing the difference in means from two independent samples (page 82). William Cochran, also a statistician at Iowa State University, joined with George for the sixth edition of the text; David F. Cox (also at Iowa) drove the work on the seventh and eighth editions, both being done after the passing of George and Bill.

In the eighth edition, they initially stick with the equal variances approach (pages 89–91), but then they later include instructions for handling the case of unequal variances, with the caveat that one ought to test for equality; if rejected, use the more complicated associated arithmetic (pages 96–98).

Contemporary Authors

Here we skip the biographical details and simply report their advice.

> **Sokal and Rohlf** (1995) was written back when folks had to do computations by hand, so a considerable portion of the text is devoted to instructions on calculations (not so necessary these days, we think). That said, they show how to do a test assuming equal variances (Calculation Box 9.6, page 225) and that assumption (Box 13.4, pages 404–405). We stand with Fisher and Snedecor and would do testing with the assertion of equal variances. They show CI construction for the case of equal variances (Box 9.6); they do not address intervals for the unequal variance case.

> **Zar** (2010), also heavy on computational details, uses the equal variance approach for testing, but discusses the unequal variance case as a violation of assumptions and gives suggestions for dealing with it (page 136). Notice that dealing with equal variances as an assumption is different from simply asserting it (à la Fisher, Snedecor, and yours truly). He then goes on to give detailed instructions for choosing between the two (pages 141–142). For estimation via CIs, he shows how to do it both ways (pages 142–145).

> **Ramsey and Schafer** (2013) use the pooled SE (page 40) so that the discussion keeps the "equal variance" flavor that will come along in regression and ANOVA. The two-independent sample case for estimating differences in means has a direct analogue when estimating proportions. For the two-sample case for estimating proportions, it is common practice these days to argue that one should pool the data to estimate the SE for the test but use independent estimates of SD for the estimation (i.e. CI) side of the business. Indeed, these authors do precisely that (page 556).

De Veaux *et al* (2013) argue that assuming equal variances adds an unnecessary assumption and so use the "don't assume…" approach for both testing and estimation. (pages 552–562). They show testing with the equal variances approach, but insist they are doing so only out of historical interest.

Utts and Heckard (2012) show in detail but then caution that the pooled approach rests on a critical assumption and so should be used cautiously (page 518).

What to Make of All This?

1. You can do whatever you want: insist on always using equal variances, or always not, or use them differentially. You can find an expert (see above) ready to back you up.

2. We still like

   a. Equal variances for testing because doing so is consistent with the null hypothesis

   b. Don't assume equal variances for the estimation part. It amounts to an unnecessary assumption.

A note… For estimation, if the variances are clearly *not* equal, then the "assume equal variances" approach is wrong, simply. If they are reasonably close in value, and you use the equal variances approach and we use the other, the resulting CIs will be almost identical. **So, the equal variance approach is either wrong or gets you the same answer anyway.**

Another note: The degrees of freedom formula under the condition of equal variances is total sample size minus 2. Without that assumption, a complicated formula is used. It will produce a smaller number. To the degree the variances differ, to that this degrees of freedom formula yields a value that is smaller than that obtained using the simpler formula.

Notes

1   There must be some reason, beyond the data, to name a value. Perhaps it is a historical value, or some threshold value of interest.

2   Very often, samples are convenience samples (whichever black bears stepped into your culvert livetrap, whoever volunteers to take your survey or participate in your exercise study), so you need to argue that the variables measured are not unduly affected by the convenience. The bears might be lighter than average, for instance, because hunger overcame their reticence to step into the trap.

3 It is easier to clarify this with a counterexample. A sample of nine (say) piglets with four from one litter and five from another would not qualify. Animals from within a litter are more likely to be like one another than animals from different litters.

4 For instance, look for other studies that (1) use similar variables and that (2) have larger sample sizes.

5 In house sparrows, males are slightly larger, on average, than females.

6 In establishing the parameter of interest, choice of order of operation (perished minus survived *versus* survived minus perished) is arbitrary. We recommend ordering them to yield a positive number. It will not affect to your conclusions, but it will make the story telling more convenient. Minus signs are confusing (to Ken, at least).

7 It is called Satterthwaite's approximation after David Satterthwaite, who derived it. You don't need to know it; your statistics package will take care of that. Note, though, that the calculation does not always yield an integer answer. That's fine, but it might surprise you the first time you see it.

8 In humans, the hippocampus plays a major role in storing short-term memories into long-term storage.

9 This simply and only to make the story easier to tell...

10 We chose 30% (and a one-tailed test) for no particular biological reason... (so let's pretend that the historical value has been 30% and we are interested in seeing whether a management intervention has reduced the mortality rate.

11 When we are being generic, we use the term "event" to correspond to "the event of interest" (which in Amy's case is death of a fish, since she is measuring the mortality rate.

12 This condition is only relevant if you are sampling from a finite population. An example might be taking a sample from the population composed of all the students in an introductory statistics class. The population size might be $N = 150$. In that case a sample of size $n = 75$ would constitute ½ of the population.

13 The "expected number" is determined by the underlying probability of a "1" and sample size. Since we don't know that probability (else we would not be sampling to estimate it), we must use our observed tallies.

14 Some authors suggest a "5/5" condition. Fine, but be sure to use the word "approximate" in your conclusions.

15 Note here that we wrote p not $\hat{p}$. We are making a test about the true proportion in the population; we will of course use the sample estimate in evaluating that test. Note also that we wrote the alternate hypothesis first. That is indeed the logical order, since the null hypothesis gets formed from it.

16 Cautionary note: This is called the "exact" method because it uses the Binomial distribution formulae itself, not some approximation. Ironically, it does not produce confidence intervals that attain *exactly* (say) the 95% confidence level. The method tends to produce intervals with a higher attained confidence level than the one you choose, implying that you are getting intervals that might be a tad wider than you need.

17 We note that this is not adding fake data. The data are the data are the data. This is simply a different way to get from your data to a confidence interval.

18 We used a standard (i.e. two-sided) interval out of habit. Strictly speaking, a one-tailed test is complemented by a one-sided interval (a lower bound in this case). People tend to be more comfortable with the standard interval; a lower

bound is more mathematically consistent. Does anybody know Tyler's phone number?

19 Anders Hald, 1998. A History of Mathematical Statistics from 1750 to 1930. Wiley, New York. 795 pages.

20 Bradley Efron, 1998. R.A. Fisher in the 21st Century. Statistical Science 13(2): 95–122. We note that Brad (inventor of the bootstrap) might also be regarded as one of the most important figures in 20th century statistics.

21 Ken added the capitalization to the word "normal" to highlight it as the name of a specific distribution, rather than a synonym of, for instance, the word "usual." We note also that currently, we would not be inclined to insert the word "Normal" in there at all, as asserting that the data come from a Normal distribution is not a requirement: the Central Limit Theorem, given a sufficiently large sample size, assures us that means, and differences in means have (at least approximately) a Normal distribution.

References

Agresti, A., and B. Coull, 1998. Approximate is better than 'Exact' for Interval Estimation of Binomial Proportions. The American Statistician, 52: 119–126.

Bockino, N.K. and D. B. Tinker. 2012. Interactions of White Pine Blister Rust and Mountain Pine Beetle in Whitebark Pine Ecosystems in the Southern Greater Yellowstone Area. Natural Areas Journal, 32(1):31–40.

De Veaux, R., P. Velleman, and D. Bock. 2013. Intro Stats (4th Edition). Pearson, New York. 815 pages.

Ramsey, F. and D. Schafer. 2013. The Statistical Sleuth (3rd Edition). Brooks/Cole, New York. 760 pages.

Schrank, A.J. and F.J. Rahel. 2004. Movement patterns in inland cutthroat trout (*Oncorhynchus clarki utah*): management and conservation implications. Canadian Journal of Fisheries and Aquatic Sciences 61: 1528–1537.

Sokal, R. and J. Rohlf. 1995. Biometry. (3rd Edition). Freeman and Company, New York. 887 pages

Utts, J. and Heckard, R. 2012. Mind on Statistics (4th Edition). Cengage, New York. 717 pages.

Zar, J. 2010. Biostatistical Analysis. (5th Edition). Pearson Prentice Hall, New Jersey. 994 pages.

15

Relative Inference for Means from Two Samples: Introducing the Bootstrap

What's novel? The whole chapter, once again. Standard statistical tools focus on arithmetic differences, but sometimes relative change is of interest. In fact, we think that it is an approach whose time has come, and this chapter is devoted to showing you how to do it for paired data.

There is a plot twist along the way. For paired data, it appeals to intuition that one could form ratios within pairs, which would measure relative change rather than use subtraction. And, indeed, you can do this, yielding a one-sample analysis on the mean of the ratios (MoR). Sometimes, though, this leads you in the wrong direction, and, instead, you should use the ratio of the means (RoM). We explain the difference and how to choose between them.

The RoM requires the use of bootstrapping for confidence intervals. So, in addition to learning about relative inference, we introduce you to elements of bootstrapping.

15.1 The Ratio of Means

We will start with two independent samples and then move on to paired data. In both cases, we will introduce the ratio of means and discuss how to make an inference using that ratio. The focus here will be on estimation via confidence intervals; testing can be done in the usual way on differences between means since $\mu_1/\mu_2 = 1$ implies that $\mu_1 - \mu_2 = 0$.

The mouse lifespan data with weekly calories set at a fixed level of 85 kcal/week, introduced in Chapter 2, were part of a larger study to examine the effects on lifespan of calorie reductions. A second group in that study had their calorie intake reduced to 50 kcal/week, with the reduction imposed after weaning. Recall that the average lifespan in that 85-kcal group was 32.7 months. The 50-kcal group averaged 42.3 months. They lived almost ten months longer, and the difference was highly significant, using a two-sample t test.

What is the importance of this result for humans? The difference in lifespans of ten months for mice doesn't translate easily to humans. How much longer did the 50 kcal group live *relative* to the 85 kcal group? That leads naturally to the statistic $\bar{Y}_{50}/\bar{Y}_{85}$. Here, we get $\bar{y}_{50}/\bar{y}_{85} = 42.3/32.7 = 1.29$. An almost 30% increase in lifespan! Holy camoly! Even if only some of that were to play out in

DOI: 10.1201/9781003609605-19

a human life, that's a lot![1] The statistic is intuitive to understand, but it comes with some issues, which may explain why you have not been taught about it in the past. We will explain the problems and then show you how to get past them using a computer-intensive method called bootstrapping.

1. Here are the problems with the RoM $\bar{Y}_1 / \bar{Y}_2$. The formula for the standard error is known to be unstable unless the denominator mean is quite precisely estimated, that is, its SE must be quite small relative to the mean itself.

2. While it is true that the sample mean from a random sample is known to be an unbiased estimator of the relevant population mean, the ratio of the two means is a biased estimator for μ_1 / μ_2. It is biased downwards slightly, especially for small sample sizes.

3. Given decent sample sizes, the Central Limit Theorem (CLT) will kick in, and so the distribution of each of the sample means will be approximately Normal. Unfortunately, the CLT does not apply to the distribution of the RoM, and so even if you could get past problem (1) and live with (2), it's not clear how to do inference. We need *some* sense of the distribution to make confidence intervals.

Let's start with (2). Given a random sample, many commonly used statistics are unbiased. The sample mean is an unbiased estimator for the population mean; ditto for a sample proportion, the difference between two means or two proportions and parameter estimates from classical regression models. What does that mean, exactly? "Unbiased" means that the statistic does not have any systematic tendency to underestimate or overestimate the target parameter. In most cultures on the planet, an accusation of being biased is not a good thing; it is something to avoid if possible. (See the Statistics/English translation dictionary in Chapter 2.) As it happens, the sample standard deviation and the sample correlation coefficient are both biased (downward) estimators for their population counterparts. Why have your teachers failed to emphasize that? The bias is slight, except for very small sample sizes, and the size of the bias gets smaller as the sample size increases. So, we rarely worry about it. The same is true for the RoM. It is indeed biased downward, but not by much. Except for very small sample sizes, it should not be a concern.

As for points (1) and (3), there is a way to generate a confidence interval that does not depend on having a formula for the SE, nor does it even depend on having any preconceived notion about the distribution of the statistic. If this sounds fantastical, it is true that when Brad Efron introduced the bootstrap (Efron, 1979), it took practitioners a bunch of years to wrap their heads around what he had created. Indeed, in 1997, when Ken used bootstrapping in a talk he gave as part of the tenure process, there was incredulity and doubt, especially among some of his senior zoologist colleagues. They were sure he was trying to pull off a "pseudoreplication" scam on them.

Any time you employ statistical inference; you are doing one of two things: generating a *p*-value from a hypothesis test or generating an estimate. Here we are discussing the latter, with the estimate being accompanied by a confidence interval. To do this, you need to make some assumptions or intelligent guess regarding the distribution of the statistic used to estimate that parameter.

Bootstrapping is the name of a simulation approach to solving this problem. Let's start by discussing bootstrapping for some chosen statistic from a single sample. Once we have that pinned down, it will be relatively simple to discuss two-sample and paired procedures. Bootstrapping is based on the idea that your data themselves are the best representation you have of the population whence it came.

15.2 The Bootstrap Algorithm

15.2.1 Conceptual Bootstrap Algorithm

The first three steps create an "empirical bootstrap distribution" of the chosen statistic[2]. The last is used to organize them for constructing a confidence interval.

1. Write down each of your *n* sample values an infinite number of times. This will generate a population of infinite size from which to do the simulation.
2. Draw a random sample of size *n* from that population. Write down the statistic.
3. Repeat (2) a large number *B* of times (10,000 will suffice and is very quickly done these days).
4. Sort the values of the computed statistic from smallest to largest.

If you were paying attention while reading this algorithm, you hopefully had a, "Wait! No way!" moment in step one. We will take care of that below, so stay with us.

Stunningly simple, except for step (1), which flat-out appears impossible! How can we trick the computer into doing bootstrapping without having to face infinity? Read on.

15.2.2 Practical Algorithm to Produce Empirical Bootstrap Samples

1. Draw a random sample of size *n* from your original sample, *sampling with replacement*, recording the mean for each bootstrapped sample. Write down the statistic.
2. Repeat (1) a large number *B* of times.

What does "sampling with replacement" mean, and why does it work? First, the "what." Suppose your five data values are: 6, 12, 17, 18, and 26 (the average is 18.8). Randomly select a number from that list. Write it down, but do not remove it from the list. Repeat. Thus, a given number is allowed to be repeated in your list. For instance, you could get 12, 6, 18, 6, and then 17, yielding an average of 11.8. As this example shows, some values will randomly *not* appear in your selection.

Why does this work? Sampling with replacement was a stroke of genius on the part of Efron. It perfectly mimics what step (2) of the conceptual algorithm does if you could possibly pull off step (1) because every time you select an observation into a bootstrap sample, it is still eligible to be picked again. Picking one value from a collection with an infinite number of each original datum leaves an infinite number of each. A simple computational trick replaces infinity!

15.2.3 Constructing Bootstrap Confidence Intervals: The Simple Percentile Method

We will write this section using 95% as a choice of confidence level. This CI procedure basically relies on two percentiles,[3] as follows:

1. Sort the values of the computed test statistic from smallest to largest.
2. Given, say, $B = 1000$ sorted values of the resampled statistic, number 26 (isolating 2.5% that are smaller) and number 975 (isolating 2.5% that are larger) in that list are the lower and upper bounds on a 95% confidence interval.

For an 80% interval, use numbers 101 and 900, isolating the outer 20% and capturing the central 80% from that list. You get the idea. It is important to note that whether you choose $B = 1000$, 10,000, or 100,000, you are still studying inference for a sample of size n. With larger values of B, the endpoints of your confidence intervals will get less wobbly, that's all. We have found that $B = 10,000$ works well for most problems.

Brad, besides being a brilliant statistician, has a delightful sense of humor. There is a statistical technique called the jackknife, which back in the 1970s was considered a computationally intense method for estimating standard errors. It worked on pretty much any statistic you cared to dream up. Hence the name, since a jackknife is considered a very flexible, multi-purpose tool.

Well. There was an 18th-century fictional character named Baron von Munchausen, whose forte was pulling off feats that were/are impossible. In one episode, he and his horse got caught in quicksand. The enterprising Baron got himself and his horse out of the quicksand by pulling them both up with his bootstraps[4]. Well, that name and idea tickled Brad. The name is also apt in that one does indeed repeatedly "tug on your data" to create the bootstrap samples.

15.3 A Better Bootstrap CI Procedure: The BCa Interval

The distribution of bootstrap estimates can be biased. One way to think of bootstrapping is as a simulation procedure that uses your original sample as the population from which the resampling takes place, and your chosen statistic as the parameter to be estimated. If the statistic you are using is biased (e.g. ratio of means, standard deviation, correlation), then the mean of the bootstrap estimates of it will be biased compared to your sample value, thereby compounding the initial bias.

Also, in many situations, the distribution of statistics is skewed. The distribution of the mean, for instance, will indeed become more and more like a Normal distribution as sample size grows, but for modest sample sizes, some skew may remain. And the distribution of other statistics (e.g. the ratio of means, which is of particular interest here) may often be skewed.

A popular bootstrap CI procedure that addresses these issues is the BCa (bias-corrected, accelerated) procedure. The resulting intervals are improved over the empirical intervals. The BCa interval requires that you estimate two parameters. The bias-correction parameter, z_0, is related to the proportion of bootstrap estimates that are less than the observed statistic. The acceleration parameter, a, is proportional to the skewness of the bootstrap distribution. These two values are used to adjust the endpoints of the empirical confidence interval.

Bootstrap procedures are showing up in statistical software, but, for instance, the ratio of means is a novel enough statistic that it might not be built into software you have access to. If you wish to program it yourself, see pages 29–36 of Manly and Navarro Alberto (2021) for details.

When applying the bootstrap to two independent samples, you independently resample the two original samples, each according to its own sample sizes. Then, for each pair of "resamples", calculate the statistic of interest. That will yield a bootstrapped distribution of your statistic.

15.4 Relative Inference for Means from Paired Data

If one is interested in *relative* change (i.e. change measured as a ratio) for paired data, one approach would be to calculate appropriate ratios within pairs and conduct a suitable one-sample analysis on the mean of those ratios (MoR). Given a decent sample size, the *t*-distribution would work. Another approach uses the ratio of the two means (RoM). Interestingly, the two do not tend to yield the same answers, and so it is worthwhile discussing that. A cartoon example is shown in Table 15.1.

TABLE 15.1

Hypothetical Paired Data to Illustrate RoM versus MoR

Pair	Group A	Group B	B - A	B/A
1	2	14	12	7
2	10	10	0	1
Means	6	12	6	4

First, notice that the mean of the arithmetic differences is the same as the difference between the two means: both equal 6. For a classical analysis, an approach based upon individual change gives the same answer as an approach based on summarized data. However, the ratio of the two means $(12/6 = 2)$ is distinctly different from the MoR, which equals 4. A detailed discussion of the reasons behind this paradox and, consequently, how to choose between them can be found in Gerow *et al* (2021). Therein are also several real examples that illustrate the consequences of using one over the other, as well as some details of relevant theory. Figure 15.1 can guide you through the choice.

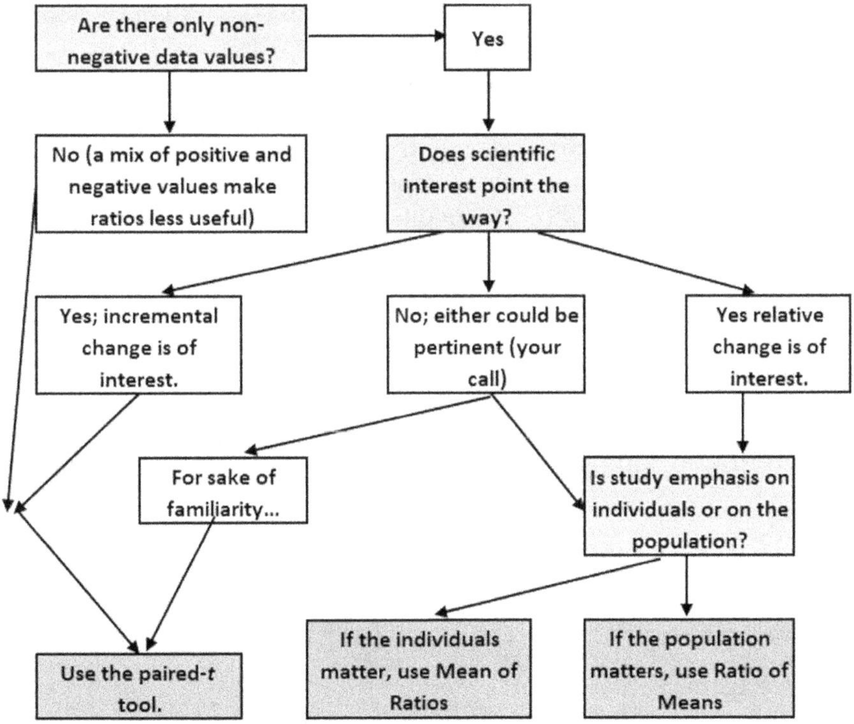

FIGURE 15.1
Key to choosing analysis methods for means from paired data.

The defining question here is whether the emphasis is on individuals or on the population itself. Some examples might help.

1. In studies of changes in percent body fat of hibernating animals, answering the question of average change of individuals makes sense. In that case, we would recommend the MoR.

2. In studies of changes in oxygen consumption rates in seals while diving as compared to resting, measuring the change on a per-seal basis makes sense, and if relative change is of interest, then the MoR is attractive.

3. Let's use the numbers from Table 15.1. Suppose Micronesia National Park is composed of two one-hectare plots, and the data show the number of some plant of interest in each plot, measured at two points in time (A, and later, B). According to the individual ratios, you might conclude that there are now 4 times as many of those plants at time B as compared to time A. But let's look at the totals for a minute. At time A, there were a total of 12 plants on the property, and then 24 by time B. The number of plants had doubled, not quadrupled. A mean is just a re-scaled total, so the ratio of means will be the same as the ratio of totals. In this situation, we don't care so much about individual plots as we do the landscape as a whole. Use the RoM.

4. In an example of monitoring fish abundance, Gerow *et al* (2021) show a stunning difference in inference based on the two methods. Yaqui chub abundance was measured in ponds in 2007 and then again in 2012. The mean of individual ratios was 56.8, suggesting almost 60 times as many chub! (This was due to one pond having only a very few in the 2007 season.) The ratio of means, on the other hand, was 2.5. This example demonstrates that the choice between the two statistics is not meaningless.

When using individual ratios (MoR), a value of 0 in the denominator will yield an undefined ratio. One can choose to replace the 0 with a judiciously chosen small value (the choice of which will affect the resulting estimate) or drop that datum. The occasional 0 will not pose such a problem for the ratio of means.

15.5 Chapter Summary

The tools (ratios of means) introduced in this chapter might be a case of "invention is the mother of necessity." Researchers have routinely used methods that focus on difference (by subtraction) in means, because that is all they were taught, and *that* is because that's all we statisticians knew how to handle easily. We believe that relative change is not a rare thing, and sometimes of more interest than additive change.

With that in mind, this chapter studied the ratio of means for two-sample and paired data. For paired data, we showed that the ratio of means is a meaningfully different parameter to estimate than the mean of ratios. If you are indeed interested in change "on average," the MoR will get you where you need to go. On the other hand, if change at the population level is of interest, you should consider the RoM.

A snag with the RoM is that one must use bootstrapping for confidence interval construction. These days, with high-speed computing, doing bootstrapping is not too difficult.

Notes

1 Luckily for us, we authors are too old for such an effect, so we won't bother ourselves by thinking about all the good things we eat and drink.
2 There are other ways to create the bootstrap sample. For instance, if you assume the data come from a population with a specified distribution, you can generate random samples from that distribution.
3 The median is an example of a percentile; it is the 50th percentile. For another example, the 10th percentile of a collection of numbers has (ideally) 10 percent of values that are smaller and 90% that are larger.
4 Interestingly, back in the days of the Baron, to say someone was trying to pull themselves up by their bootstraps, was a clear insult. Despite the obvious impossibility of the idea, we now use the phrase as a form of praise.

References

Efron, Brad. 1979. Bootstrap methods: another look at the jackknife. Annals of Statistics 7(1):1–26.

Gerow, K., D.R. Stewart, and C. Farris. 2021. Relative inference for paired data: more than meets the eye. Bulletin of the Ecological Society of America 103(2): 1–12.

Manly, B.F.J. and Navarro Alberto (2021) Randomization, Bootstrap and Monte Carlo Methods in Biology. (4th Edition). CRC Press.

16

A Brief Introduction to ANOVA

Our approach to data analysis is mostly regression driven. So, there will not be multiple chapters on different designs in the Analysis of Variance (ANOVA) world: one-way, two-way, block designs, split plot, Latin squares, and so on. Instead, we will discuss some of the terms used by ANOVA-speakers, and then show that regression and ANOVA models are, at their core, the same analyses, but with different presentation styles. They all fall into the family of "linear models"; see section 8.1 for a discussion of that term. When the two methods were created, their synonymity was unknown, and so separate cultures of usage and language developed, and we provide a brief ANOVA-regression translation dictionary to address that.

There are two topics we spend some time on. One is "contrasts among treatments." This is something you might not need (so feel free to skip it), but it is an approach that is not overly difficult to implement and not readily available in many statistics packages. We will show you how... The other topic is a brief look at the so-called multiple comparisons problem. We introduce two methods for addressing the problem. One is the Bonferroni method (and we include a creative variation that you might consider); the other is a more contemporary method called the false discovery rate (FDR) method. While theoretically deep, it is not difficult to implement. If you have many comparisons, it is a more powerful method (i.e. is more likely to find effects) than the Bonferroni method.

16.1 ANOVA and Regression are Close Cousins

ANOVA is the acronym used for methods that were developed for examining arithmetic differences in means for studies with different treatment groups[1]. First, we want to show you that regression and ANOVA are different flavors of the same underlying tool, namely a linear model. Whether you choose to follow the ANOVA line of analyses or the regression line is, to some extent, a matter of taste and (more likely) how you were trained. There are a couple of ANOVA-related topics that we will spend some time on. One of these is the idea of a contrast among treatment means. The other is the problem of multiple comparisons. There, we will discuss the Bonferroni approach as well as a modern method that is particularly useful when you have many comparisons to make, namely the FDR method.

DOI: 10.1201/9781003609605-20

The very first applied statistics text came to us from Ronald Fisher in 1925 (Fisher, 1938, 7th edition). Back then, computational equipment had three primary components: a pad of paper, a pencil, and an eraser. Needless to say,[2] the text is chock-full of arithmetic and careful instructions on how to lay out your data on that piece of paper in rows and columns: do this or that to the numbers in each column, write the answer at the bottom of the column, etcetera. The book introduces both regression and ANOVA as distinct methods. Indeed, one would never know by looking at the equations that they had anything in common. Fast forward to the end of the 20th century. Now we know them to be variations on a common idea, namely a linear model.

You might find the following short discussion of regression and ANOVA terminology useful. Unlike the statistics/English dictionary at the end of Chapter 2, this one presents synonyms rather than different definitions of the same words. We are native "regression speakers," so for the fun of it, we will write this as though you are an ANOVA-taught practitioner.

Regression was designed for a **numerical predictor**. Initially, it was for just a single predictor; we would now call such a model a simple linear regression model. Imagine, on the other hand, a study with three diets. One is a standard diet, the other two are altered somehow, and some health-related outcome measure is the response. Or a study on concrete strength with three different types of aggregate being considered. The "thing" being studied (diet, type of aggregate) is called a **factor**, and the various flavors of the factor are called different **levels** of the factor. Some of this terminology arose due to early studies in agriculture, where, literally, different levels of fertilizer might be under consideration.

Imagine now a study with different diets; the goal is to see which is most effective in the growth rates of pigs being raised for meat. The weight of these piglets at birth would likely have some effect on the outcome independent of diet, so at some point, ANOVA technology was developed to include a numerical predictor, which the developers of ANOVA named a **covariate**. There is our first pair of synonyms. We say **numerical predictor**; you say **covariate**. Same thing, different name.

Just as ANOVA developers had to learn how to accommodate covariates, eventually regression developers realized that including variables such as sex or species would be very useful. Accordingly, **categorical variables**, each with some number of **categories**, were introduced. Two more pairs of synonyms. **Factor** and **categorical variables**, each with **levels** or **categories**. In ANOVA-speak, a treatment group can be formed from a specific combination of factor levels. The example in section 16.2 exemplifies this. Categorical predictors are often deployed using indicator variables, composed of zeros and ones. See section 10.2.1 for a more detailed introduction. A standard summary statistic from regression models has always been R^2: the proportion[3] of variation in the response that is associated with the model. In recognition of the equivalence of regression and ANOVA, that summary statistic has become a regular feature of ANOVA outputs in many statistics packages.

TABLE 16.1

An Illustration of a One-Way Layout
for a One-Factor ANOVA

Factor Level	A	B	C
Data	7	6	2
	9	12	5
	11	15	4
	4	10	7
	8	9	6

Most regression analyses these days are multiple regression; models with two or more predictors, some of which are numerical, some categorical. Back in the early days (pencil and paper computing technology), the instructions for doing ANOVA were different depending on how many factors there were, among other features. The differences were sufficiently profound that there could be many chapters in a text, each devoted to just one design. And each had its own name. Let's look to see why.

Suppose you have a single factor with three levels (A, B, and C). The first step would be to lay out your data as shown in Table 16.1.

Then the instructions would say to add up all the values in each column and write the answer below the columns. Then divide each total by the number of values to obtain the sample mean and write *that* down in a row below the totals. And on and on. This layout was called a one-way layout (organized only by columns), and the resulting analysis was dubbed a **one-way ANOVA**. If you have two factors, you can imagine a similar table, but with rows labeled according to the levels of the second factor. The data values for each combination would be listed in the relevant cells of the table. Yup. You got it: a **two-way ANOVA**. And on and on: three-way, block designs, split-plot designs, and so on. Each merited its own standalone chapter because the instructions for doing the arithmetic were unique to the design.

In that same spirit, analyses that included a covariate were sufficiently different computationally that they merited their own chapter. This variation came to be dubbed ANCOVA (Analysis of Covariance). The covariate might be a nuisance variable (as in the foregoing pig weight example) that needs to be accounted for, but sometimes it is important in and of itself.

16.2 Contrasts: Inference Using Combinations of Means

It is often of interest to make comparisons among specific treatments. For example, you might want to compare the average of group A to that of group B. Or perhaps comparing A to the average of B and C is of interest. If your interest is only in so-called pairwise comparisons (one mean compared to one other), then you needn't bother with this section at all. A pairwise

TABLE 16.2

Mean Phospholipid Levels for the Four Treatment Groups in the Lamb Study

Treatment	Morning, Control	Morning, Estrogen	Afternoon, Control	Afternoon, Estrogen
Label	MC	ME	AC	AE
Average	13.28	19.36	36.53	27.81

comparison can be handled by a two-sample t or a paired t (in the case of a block design). Simply set the data from the two treatments of interest into your favorite statistics package, and off you go. Otherwise, read on.

Fair warning: this section is somewhat tedious, and you are not overly likely to need it. So, you might ask: why? For one thing, if you *do* want to make comparisons that are more complicated than simply comparing two means, most statistics packages don't have that feature built in, and it is fairly easy to set them up in Excel.

As part of a larger study, Wilkinson *et al* (1954) studied the effect of estrogen on blood plasma phospholipids in lambs (Steel and Torrie, 1980, page 343). Two factors were studied: time of day (blood drawn in the morning or in the afternoon) and presence or absence of estrogen (implanted subcutaneously in the ear of the lamb). There were five animals in each treatment group. Treatment means are shown in Table 16.2.

Glancing at this table, it appears that the effect of estrogen might differ in the morning (an increase in phospholipids) from the afternoon (a decrease). Before continuing, we note that the analysis supported an interaction between the two treatments. With that included, model assumptions were reasonably met. Here is the ANOVA table output (Table 16.3):

TABLE 16.3

ANOVA Table Output for the Phospholipid Study

Analysis of Variance

Source	DF	Adj SS	Adj MS	F-Value	*p*-Value
Pm	1	1256.75	1256.75	52.93	0.000
Trt	1	8.71	8.71	0.37	0.553
Pm*Trt	1	273.95	273.95	11.54	0.004
Error	16	379.92	23.75		
Total	19	1919.33			

Notes:

[1.] We coded the time of day as 0 for morning, 1 for afternoon. As is our wont, we named the factor by the level coded as a 1. Ditto for treatments: 0 for control, 1 for estrogen-treated.

[2.] It appears at first glance that there is no effect of estrogen ($p = 0.553$). This is a false impression, which we will discuss below. The short version: if you have an interaction between two factors, you must keep the individual factors in the model as well, without regard to apparent lack of significance.

We won't claim any special knowledge in choosing the following comparisons, but suppose you are interested in

1. The effect of estrogen in the afternoon. For this, we can subtract the average of AE from the average of AC[4]: $\bar{y}_{AC} - \bar{y}_{AE} = 8.72$. This question can be addressed by a simple two-sample t-test, a point we will return to below.
2. The effect of estrogen. For this, we can use the average of MC and AC and contrast that with the average of ME and AE: $\frac{1}{2}(\bar{y}_{MC} + \bar{y}_{AC}) - \frac{1}{2}(\bar{y}_{ME} + \bar{y}_{AE}) = 24.9 - 23.6 = 1.3$.
3. Whether the effect of estrogen is different in the morning versus the afternoon. Here we will measure the effect of estrogen in the afternoon and in the morning, and subtract them: $\frac{1}{2}(\bar{y}_{AE} - \bar{y}_{AC}) - \frac{1}{2}(\bar{y}_{ME} - \bar{y}_{MC}) = \frac{1}{2}(-8.7 - 6.0) = -7.4$.

These contrasts were chosen, not for their interest biologically, but because they all, at first glance, seem quite different in structure. It is our goal here to show their commonality, that they are all, in fact, variations on a theme.

Labels	$\bar{Y}_1 : \bar{Y}_{MC}$	$\bar{Y}_2 : \bar{Y}_{ME}$	$\bar{Y}_3 : \bar{Y}_{AC}$	$\bar{Y}_4 : \bar{Y}_{AE}$
Means	13.3	19.4	36.5	27.8
		Coefficients		
Contrast 1	0	0	1	−1
Contrast 2	0.5	−0.5	0.5	−0.5
Contrast 3	0.5	−0.5	−0.5	0.5

There are certain requirements of the coefficients in each row for them to be part of a contrast:

1. Every mean gets a coefficient, even if it is a zero.
2. The coefficients that are positive and those that are negative must each add to one.
3. Collectively, the coefficients.
4. In a contrast must sum to zero.

It is arbitrary which "team" gets the negative coefficients, and which the positive. We tend to choose them so the answers are positive numbers. We broke the "rule" for the third contrast, the why of which we will return to below.

It might seem like an unnecessary abstraction to write them like this, but keep reading. The payoff will come. Each contrast is a particular case of the following: $c_1\bar{Y}_1 + c_2\bar{Y}_2 + c_3\bar{Y}_3 + c_4\bar{Y}_4 = \Sigma_{i=1}^4 c_i\bar{Y}_i$, where the coefficients c_i are chosen by the researcher, subject to the three conditions above. We can do a t-test for each of these using the standard deviation among the residuals (s_{res}) and the degrees of freedom (df) from the error line of the ANOVA table. The MSE (mean square error) in the ANOVA table is the estimated variance among the residuals, so $s_{res} = \sqrt{23.75} = 4.87$, and $df = 16$. The standard error of a given contrast is $s_{res}\sqrt{\sum_{i=1}^{4} \frac{c_i^2}{n_i}}$. If the sample sizes are all equal, and they are here, this formula simplifies to $\frac{s_{res}}{\sqrt{n}}\sqrt{\sum_{i=1}^{4} c_i^2}$, where n symbolizes the common sample size. For our three contrasts, we have

1. $SE = \frac{4.87}{\sqrt{5}}\sqrt{\sum_{i=1}^{4} c_i^2} = 2.18\sqrt{0^2 + 0^2 + 1^2 + (-1)^2} = 2.18\sqrt{2} = 3.08.$

2. $SE = \frac{4.87}{\sqrt{5}}\sqrt{\sum_{i=1}^{4} c_i^2} = 2.18\sqrt{0.5^2 + (-0.5)^2 + 0.5^2 + (-0.5)^2} = 2.18\sqrt{1} = 2.18.$

3. Same as (2), since the coefficients are all the same size...

Now you can set up the t-statistic for each and readily look up the p-value for the test of significance. Here, $t = \frac{C}{SE}$, where we used C to symbolize the measured contrast. So...

1. $t = \frac{8.72}{3.08} = 2.83$, and a t lookup with $df = 16$ yields $p = 0.012$.
2. $t = \frac{1.32}{2.18} = 0.61$, yielding $p = 0.55$.
3. $t = \frac{-7.4}{2.18} = -3.40$, and $p = 0.004$.

Conclusions:

1. Blood plasma values are higher in the control group than the estrogen-treated group by 8.72 in the afternoon, and the effect is significant, against the usual $\alpha = 0.05$. Perhaps not Nobel-prize significant, but at least it appears to be real.

2. Without taking time of day into account, estrogen appears to have no significant effect.

3. The effect of estrogen is strikingly different from morning to afternoon. By peering into the calculations for this contrast, there appears to be a positive effect in the morning (significant? We don't know yet), and a significant negative effect in the afternoon (contrast 2).

These three conclusions tell an apparently confusing story. That being said, here is a short list of the highlights:

1. Lipid levels are significantly higher in the afternoon than in the morning,

2. Estrogen appears to have little effect on lipid levels in the morning, but

3. It has a stout depressing effect in the afternoon.

16.3 The Problem of Multiple Comparisons

There are many flavors of multiple comparisons (pairwise comparisons among all treatments, all other treatments against a standard, seeking to find the best among a group of treatments, and so on). We will focus here on situations and methods that are appropriate for data sets we have run into over our careers. At some points in writing this text, we have had to wrestle with FOLO[5]; this is one such point. Designed experiments have not been a major feature in our experience, and so we will inevitably leave out ideas that might indeed be useful to some.

We will discuss the Bonferroni method, suitable for when the number of comparisons is modest. There is a modern approach, the FDR, that is quite useful for cases where you might have a very large number of tests done simultaneously; certain types of genomic testing might easily have more than 100!

16.3.1 The Bonferroni Method

Imagine that you have an experiment with five treatment groups, and that you want to make all possible comparisons (there are 10 of them). Suppose further that the general null hypothesis (none of the treatments differ from one another in their effect on the response) is true, and that you do the tests using good old $\alpha = 0.05$. What is the chance of one or more "false significance" conclusions? The phrasing of the question, while legitimate, is a pointer to a minor headache. You could, for instance, have one false significance (which could be any of the tests one through ten), or you could have two (test one and two, one and three..., nine and ten; there are 45 of these) or... we won't go on.

Easier is to see the probability of at least one false significance as one minus the probability of zero such happenings. The key to it is that, given $\alpha = 0.05$, and that the null happens to be true, the probability of *not* falsely rejecting the null is $1 - \alpha = 0.95$. The chance of getting away with this with two (independent) tests is 0.95^2; you can see how this is going to go: $\Pr(1, 2, 3, ..., \text{or } 10) = 1 - \Pr(0) = 1 - 0.95^{10} = 1 - 0.6 = 0.4$. It is pretty close to a coin flip that there will be at least one false significance. This, in a nutshell, is the so-called problem of multiple comparisons. These ten tests are not completely independent of one another, but the general idea holds.

We are going to play games with alpha here, and we will need (at least) two. One will be the alpha level we use on each test (α_T), and the other will be the false significance rate across the entire family of tests (α_F). In the example above, $\alpha_T = 0.05$, and $\alpha_F = 0.40$. The simple Bonferroni method says to use $\alpha_T = \frac{\alpha_F}{T}$, where α_F is your choice for the family-wise rate, and T is the number of tests. In our example, if we choose $\alpha_F = 0.05$, we get $\alpha_T = \frac{0.05}{10} = 0.005$. If you do the math, you will see that $\alpha_F = 0.049$, very close to the target value.

If you have experience doing analyses, this approach might make you squirm because you know that only a very small p-value will allow you to declare a significant effect. That, in turn, means your risk of missing an effect is a little higher. In fact, a legitimate complaint about the Bonferroni method is that it can be somewhat conservative (and more and more so as the number of incorporated tests goes up). For instance, with 20 tests and $\alpha_F = 0.05$, $\alpha_T = 0.0025$.

What to do? There are options. You might simply focus on the comparisons of high interest and leave off the others. You might consider using a larger α_F. If you are willing to live with, say, $\alpha_F = 0.10$, then the Bonferroni method for our example yields $\alpha_T = 0.01$.

The procedure suggested above we will now call the "classical Bonferroni." There is a creative variation that might be useful to you. What makes the method work is not that each α_T must be equal, but they sum up to α_F. Suppose in the foregoing example that two of the questions are of primary interest, and the others quite less so. You could, for instance, use 0.01 for those two and divide the remaining 0.03 among the remaining questions. A choice with consequences, but a fair choice.

16.3.2 The False Discovery Rate

The Bonferroni method of adjusting alpha for a series of multiple comparisons has the well-known unfortunate property of being unduly conservative. If you have many comparisons, it becomes exceedingly difficult to declare any of them as significant.

Yoav Benjamini and Yosef Hochberg published a groundbreaking paper in 1995 that presented a method that is quite less conservative than the Bonferroni, a method they named the FDR. The name is a bit unfortunate, since alpha is also a "false discovery rate." The FDR is defined as the proportion of false significances you are willing to live with among a list of declared significant findings. Alpha, on the other hand, is defined as the proportion of false significances you are willing to live with among a list of tests for which the null hypothesis is true in every case. The FDR definition does not include "the null hypotheses are all true." This is most likely a more realistic starting point, in any event.

The FDR yields more significant findings than does the Bonferroni, given a choice for FDR that is equal to the chosen family-wise error rate in the Bonferroni procedure.

Their method has become quite useful, especially in disciplines where there may legitimately be over 100 tests to be performed. For example,[6] genome-wide association studies seek to identify which (if any) of millions of sites in a DNA alignment are associated with organismal traits by testing each site against the trait data individually. Genome environment association methods are similar but seek to identify if environmental variation is associated with genomic variation, again by testing this association at each site of the genome. Similarly, when testing for differential expression of RNA sequence data, tests are made at thousands of genes to see if they differ between two or more experimental conditions, again requiring one to account for the many comparisons that are being run at the same time between the same experimental samples.

The theory behind the method is quite deep. Fortunately, the application follows a straightforward recipe. We will share that with you now and then illustrate the method with a small example.

16.3.2.1 FDR Method

1. Pick a satisfactory FDR; we will label that here as ϕ to differentiate it from α.
2. For T tests, order the p-values from smallest $(p_{(1)})$ to largest $(p_{(T)})$.
3. Compare each $p_{(k)}$ ($k = 1, 2, \ldots, T$) to $k\frac{\phi}{T}$ (see note below).
4. Find the largest k such that $p_{(k)} < k\frac{\phi}{T}$. The tests associated with $p_{(1)}$ through $p_{(K)}$ are declared significant.

Note: For example, with $\phi = 0.05$, and $T = 10$, the smallest p-value gets compared to $\phi / 10 = .005$, the second smallest to 0.010, then 0.015, and so on. The largest gets compared to 0.05 itself. Notice that the smallest value is in fact the Bonferroni α_T.

Example

This example comes from real data, a study with six treatment groups, yielding 15 possible pairs of comparisons. The details are unimportant here; we truly just want to compare the performance of the two methods. Many of the differences were sufficiently profound that their p-values were very small. With 15 tests, and (for the sake of comparison) $\phi = \alpha_F = 0.05$, and the Bonferroni alpha is $\alpha_T = 0.0033$. The first seven tests in Table 16.4 have p-values less than this and so would be declared significant via the Bonferroni method. The FDR approach declares the first 13 of them to be significant. The FDR approach is more powerful (i.e. it enables you to find more effects), which will be particularly important in any work like the genomics setting described above.

16.4 Chapter Summary

In this chapter, we discussed the deeper connections between ANOVA and regression. Simply put, they are different examples of linear models. We showed how to form contrasts among treatment means and introduced the Bonferroni and FDR methods for multiple comparisons.

TABLE 16.4

Illustration of FDR Approach for
a Study with 16 Comparison

Test	*p*-Value	FDR Value
1	0.00000	0.0033
2	0.00000	0.0067
3	0.00000	0.01
4	0.00000	0.0133
5	0.00001	0.0167
6	0.00011	0.02
7	0.00242	0.0233
8	0.00341	0.0267
9	0.00466	0.03
10	0.00982	0.0333
11	0.0117	0.0367
12	0.017	0.04
13	0.0294	0.0433
14	0.073	0.0467
15	0.622	0.05

The FDR approach is quite less conservative than the Bonferroni, especially for cases with many simultaneous comparisons.

Notes

1 The name ANOVA made sense 100 years ago, but if statistics were being invented now, we think we would not have a method that goes by that name because it is the goal of every statistical analysis to analyze variation, seeking to identify sources of that variation.
2 But here we go, saying it anyway…
3 It takes on values between 0 and 1 but is usually reported as a percentage.
4 We tend to order our comparisons to yield positive numbers, to simplify the discussion.
  If A – B = 10, then B – A = –10. Same thing, except that the first form is easier to discuss.
5 You might be familiar with FOMO (the Fear of Missing Out); here, it is the academic's Fear of Leaving Out something that someone else might deem important. Such is life.
6 Thanks to Sean Harrington and Nic Blouin of the University of Wyoming INBRE Data Science Core for these examples.

References

Benjamini, Y., and Y. Hochberg. 1995. Controlling the false discovery rate: a practical and powerful approach to multiple testing. Journal of the Royal Statistical Society. Series B (Methodological), 57(1): 289–300

Steel, R.G.D. and J.H. Torrie. 1980. Principles and Procedures of Statistics: A Biometrical Approach. (2nd Edition). McGraw-Hill, New York. 633 pages.

Wilkinson, W.S., A.L. Pope, P.H. Phillips, and L.E. Casida. 1954. The influence of diethylstilbestrol on certain blood and liver constituents of lambs. Journal of Animal Science 13(3): 684–693.

17

Response Feature Analysis for Repeated Measures

17.1 Introduction

Some studies have data repeatedly measured on the same unit (e.g. a person, plot, or reach of a stream). Often, the basis for the repeating is time, but not always. A classical approach to such data (repeated measures analysis) is not something one typically gets exposed to in one or two graduate classes in applied statistics. As it happens, there is an approach to such data, a "summarize, then analyze" approach we call a response feature analysis (RFA), that simultaneously quite simplifies analyses and allows the researchers to answer a broader array of questions than a straightforward repeated measures analysis can handle.

The first step requires a separate summary of the values across the repeats, separately for each unit. Creative choices of summaries are what give the approach its scientific power. We will illustrate the RFA with two fisheries examples (for one of which, the repeats are not across time) and an interesting wildlife example. In this latter one, a very creative summary leads to simple analyses answering questions of scientific interest.

Repeated measures data may arise in a designed experiment where, for instance, one group of subjects gets exercise coaching, and the other group is a control group. In that case, some measure of fitness or activity levels, measured over time, might be the response. In an ecological setting, the abundance of some species of interest might be monitored over time. While time is often the factor of interest, one example below exposes individual fish to different stream flow rates (in a lab study with aquaria) to study the impact of flow rate on their ability to catch food floating by. Repeated measures analyses sometimes pose a challenge for researchers in that the methodology is not one you typically learn about in one or two graduate-level statistics classes.

There is a lovely two-stage alternative to classical repeated measures. It sometimes goes by "summarize, then analyze" (e.g. Ramsey and Shafer, 2002); we call it a RFA after Everitt (1995; 2002, page 91) and Everitt and Pickles (2000). See also Gerow et al (2021). The term RFA is not very descriptive (or prescriptive) of precisely what you should do; as you will see from the examples here, that is in fact precisely the point. You choose some creative way to

DOI: 10.1201/9781003609605-21

summarize the data separately for each experimental or sampling unit and follow up with a usually simple statistical analysis.

In fact, this idea was introduced to you very early in your statistics education: a classical paired t analysis proceeds by subtracting values within pairs (the summary stage), followed by a one-sample analysis of the differences. Unfortunately, the creative aspect of that first stage does not get taught. For example, if you are interested in relative change, you could calculate ratios within pairs instead of arithmetic differences; see Chapter 15. We hope the examples here will show you that RFA can be a very powerful and flexible analytic tool. Using this approach exposes the underlying structure of the data, which in turn facilitates and clarifies assessment of necessary assumptions, which might get lost in more complex analyses (Murtaugh, 2007). Most importantly, RFA may enable you to discover and answer questions that a classical repeated measures analysis simply cannot address. RFA also may aid researchers in recognizing that complex statistical analyses are not always necessary to answer complex research questions (Murtaugh 2007).

17.2 Examples

17.2.1 Example 1

This example involves a typical repeated measures data set with permanent study sites repeatedly sampled over time. The data come from Alcova Reservoir in Wyoming and are part of a larger study quantifying changes in the abundance of Rainbow Trout (*Oncorhynchus mykiss*) over time (data source: Dan Yule). The data indicated that Rainbow Trout abundance was increasing from 1994 to 1998 (Figure 17.1a). These data display a pattern of curvilinearity, suggesting that the change in abundance per year is relative (i.e. a percentage of the previous year's abundance), rather than a constant increment every year. Moreover, as is common with biological data, the variation increases with mean abundance (suggesting that the variation is also proportional). As such, the data provide an opportunity for RFA: regression across time, with the response log-transformed (Figure 17.1b). A simple linear regression was used to fit the data using log-transformed abundances as the response variable.

Focusing on rates of increase, the seven slopes are 0.054, 0.063, 0.062, 0.074, 0.006, 0.054, 0.108, which back transform to 1.13, 1.16, 1.15, 1.19, 1.01, 1.13, and 1.28. The mean of the back-transformed slopes is 1.15 (95% CI = (1.08, 1.22)). That is, we estimate a 15% yearly increase and, using a one-sample t, we are 95% confident that the true rate is between 8% and 22%. Given the low sample size, one might be skeptical of the validity of using the t distribution. Accordingly, we generated a bootstrapped confidence interval yielding

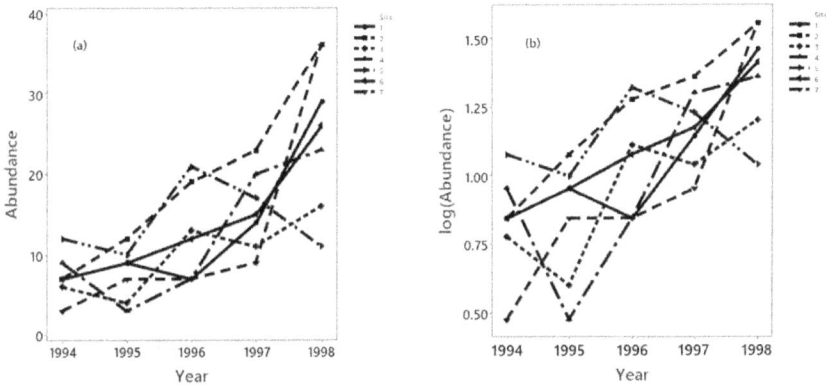

FIGURE 17.1
Fish abundance (a) from Alcova reservoir across time and log-transformed abundance (b).

the interval (1.09, 1.21). The bootstrap and the *t*-based intervals are in good accord. The fact of the intervals being quite similar is circumstantial evidence that, in fact, the Central Limit Theorem has kicked in and the *t*-based approach is likely fine. It is important to note here that the summary stage uses a statistical model (simple linear regression) to generate the measurement of interest for each location. It is, however, not a statistical analysis *per se*. We are not interested in, for instance, whether an observed slope of 0.054 is statistically significant; we only need that the observed measurement is a reasonable one for the question being asked. The slope from a simple linear regression on the raw data would not have been as credible since the relationships are visibly curved.

17.2.2 Example 2

This example of RFA demonstrates another degree of creativity in the summary stage. In short, a logistic regression is used on each fish to measure, for that fish, how the odds of catching prey diminish as stream velocity increases. Again, the summary stage uses a statistical tool to obtain one measurement of interest for each individual fish, but the result is just that: a chosen measurement, not the end point of a statistical analysis.

Bozeman and Grossman (2019) conducted feeding trials on a drift-feeding stream fish (Arctic Grayling, *Thymallus arcticus*) to determine whether stream velocity affected prey capture success. This is an example where the "repeating factor" is not time but rather stream velocity. Individual fish were subjected to seven increasing stream velocities in an experimental stream flume. Specimens were fed nine prey items at each treatment velocity, and the number of prey captured and missed at each velocity was recorded. We plotted the proportion captured for each fish at each velocity (Figure 17.2a)

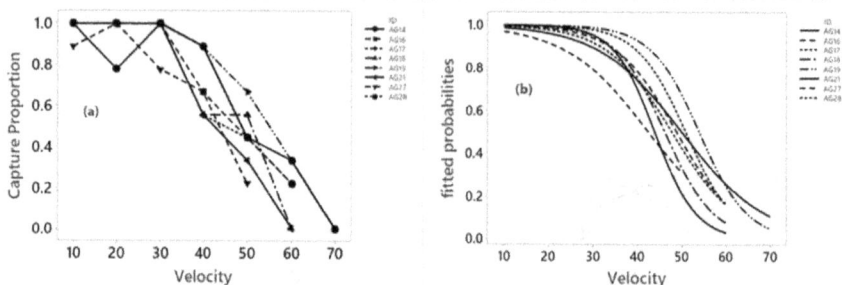

FIGURE 17.2
Observed capture probabilities (a) and fitted probabilities using logistic regression (b) for each fish.

for a subset of the fish (a subset only, to have graphs that are not too cluttered). We then fitted a logistic regression separately to each fish, estimating capture probabilities as a function of velocity (Figure 17.2b). This was a prospective study because capture probabilities were an observed property of the data; that is, they were random and not pre-chosen. In such a case, it is legitimate to report and discuss probabilities. See Chapter 13 on logistic regression for more details.

For each fish, we used the odds ratio as a measure of how the odds of prey capture change with increased velocity. From our sample of 15 fish, the calculated mean odds ratio is 0.82; the SE = 0.014, 95% CI = (0.78, 0.85). It is easier to interpret this ratio by reporting the size of the decline in the odds (and switching to percentages): in short, the odds of prey capture drop by 18%; SE = 1.4%, 95% CI = (15%, 21%), for each 10 cm/s increase in treatment velocity.

17.2.3 Example 3

This example truly illustrates the creativity that can be so useful in the summary stage of an RFA. Justin Clapp, then a graduate student at the University of Wyoming working with Jeff Beck, collected massive amounts of location data on bighorn sheep translocated into wildlands that were new to them (Clapp *et al.*, 2014). Using collared sheep and GPS technology, there were multiple locations per day, taken on each animal. This was done for many months. What Justin was interested in was the amount of movement over time, based on those location snapshots. When Justin first showed Ken a time series graph of the movement data and asked him how he thought they should be analyzed, he had no idea. But he also had no idea why they collected the data in the first place. Perhaps that would suggest a question to ask…

They were interested in estimating "time until settling." How long did it take for these animals to feel like they were at home? How would you know? When they stop moving so much. Ah, a clue. The animals were relocated around the turn of the year. Justin told Ken that they all had for sure settled by the following summer. That led to the following summary approach.

He used the standard deviation of weekly movement data taken in summer as a measure of how much movement to expect from a settled animal. Then, for each animal, the SD in movements on a weekly basis (from early after relocation) was recorded, and when the SD fell (and remained) below a value just above that "settled SD" level, the animal was deemed to have settled, and the time when that happened was duly recorded. That reduced the data to one measured value per animal (time until settling), and greatly facilitated subsequent analyses, comparing, for instance, settling times for ewes to those for rams. If you are curious about that question, we recommend Clapp *et al* (2014).

17.3 Chapter Summary

The response feature approach to repeated measures data can answer a broader suite of questions than can typically be done with classical repeated measures methods. This increase in scientific power does not come with a "More complicated analyses are required" price tag. Somewhat paradoxically, the resulting statistical analysis is usually quite simpler than the methods required for a usual repeated measures analysis.

References

Bozeman, Bryan B., and Gary D. Grossman (2019). Mechanics of foraging success and optimal microhabitat selection in Alaskan Arctic Grayling (*Thymallus arcticus*). Canadian Journal of Fisheries and Aquatic Sciences 76(5): 815–830, https://doi.org/10.1139/cjfas-2018-0115.

Clapp, Justin G., J.L. Beck and K.G. Gerow. 2014. Post-release acclimation of low-elevation, non-migratory bighorn sheep. Wildlife Society Bulletin 38(3): 657–663

Everitt, Brian S. 1995. The analysis of repeated measures: a practical review with examples. The Statistician 44(1): 113–135.

Everitt, Brian S., and A. Pickles. 2000. Statistical Aspects of the Design and Analysis of Clinical Trials. ICP, London

Everitt, Brian S. 2002. "Analyzing Longitudinal Data: Beating the Blues", Chapter 6 in "A Handbook of Statistical Analyses Using S-Plus. (2nd Edition). Chapman and Hall/CRC. 240 pages.

Gerow, K., B. Bozeman, and G.D. Grossman. 2021. Response feature analysis for repeated measures in ecological research. Bulletin of the Geological Society of America 00(00): e01866. https://doi.org/10.1002/bes2.1866

Murtaugh, P.A. 2007. Simplicity and complexity in ecological data analysis. Ecology 88(1): 56–62.

Ramsey, F. and D. Schafer. 2002. The Statistical Sleuth. (2nd Edition). Duxbury. 742 pages.

18

Epilogue

This book rose out of class notes that both Ken and Jorge had developed over their respective careers teaching statistics to (mostly) young researchers. These people are hardworking, intelligent, and passionate about their chosen disciplines. Some of them happen to be innately mathematically minded, but not most. Over the years, puzzled frowns, persistent questions, and many fruitful discussions guided and inspired us to develop approaches to the subject that do not overly lean on mathematics. We had the luxury of working in the late 20th and early 21st centuries, which meant that if we could find a conceptual path to understanding, computers could do the number crunching.

Multiple regression models, in various guises, are core tools for many researchers, and so this book focuses on them, with an introduction to various elements in the simpler setting of regression modeling with a single predictor. We illustrate the (not unlimited, to be sure) power of the Central Limit Theorem and its role in supporting classical methods. Principles and procedures of hypothesis testing and estimation via confidence intervals for the other two legs of the foundation of classical frequentist methodology.

The classical methods were developed to make inference on arithmetic differences in means (i.e. by subtraction). This is in part due to the tools that were available for statistical work in the early decades of the 20th century; pencil and paper (and a large eraser) had to do. We believe that biological effects and relationships are often relative in nature. Being able to address analyses from that perspective could be helpful to researchers, so we devote two chapters to relative inference: Chapter 9 on the role of logarithmic transformations in models and Chapter 15 on relative change for means in two-sample settings.

While we succeeded in mostly keeping our attention on essentials, we were unable to resist a few "bonus topics" (in addition to relativity) that we think could be useful. Ratio regression is a regression model that arises naturally for the case where the response variable is proportional to the values of the predictor. We think this model deserves more attention. Logistic regression doesn't usually make it into the curriculum of a first graduate applied statistics course, for good reason. We felt that a conceptual introduction could serve researchers well, since they are likely to run into the method in readings in their discipline. In Chapter 17, we introduced an approach to repeated measures data, namely a response feature analysis. It is a two-step procedure: first comes a summary of the data for each subject, with the specific summary

DOI: 10.1201/9781003609605-22

designed to get at a question of interest. What follows is often a one-sample analysis of the summary values. It is scientifically powerful in that it allows researchers to address questions that often cannot be tackled via standard repeated measures approaches. It has a bonus feature that the resulting statistical analysis is often quite simple, and within the grasp of someone comfortable with the essentials.

The topics we covered in most of the chapters are undoubtedly similar to other sources on applied statistics, and these statistical methods might be described in much greater detail in other applied statistics books, particularly regarding the equations and algorithms. However, we firmly believe that the colloquial "first-person" voice makes this book useful for researchers regarding the practical application of statistical analysis from their perspective.

Index

Note: Page references in *italics* denote figures, in **bold** tables and with "n" endnotes.

For Product Safety Concerns and Information please contact our EU
representative GPSR@taylorandfrancis.com
Taylor & Francis Verlag GmbH, Kaufingerstraße 24, 80331 München, Germany